BUILDING YOUR BARNDOMINIUM

A Comprehensive Guide to DIY Barn Home Building, Creating Rural Living Home Designs and Unique Barndo Spaces on Any Budget

By

BarnDream Builders

Table of Contents

Mary’s Barndominium

In the heart of a quiet countryside, where the rolling hills met the endless sky, Mary stood at the window of her brand new dream barndominium. The structure, a harmonious blend of rustic charm and modern comfort, was a manifestation of her unwavering spirit and the culmination of years of dreams and hard work.

Mary's journey had been one of resilience and determination. Raised in the city's concrete embrace, she longed for a connection to nature, a retreat from the cacophony of urban life. With a heart full of dreams, she took a leap of faith, leaving behind the familiar streets for the promise of open meadows and clear skies.

The barndominium, with its weathered barn exterior concealing a contemporary interior, symbolized Mary's fusion of past and present. The large windows, strategically placed to capture the essence of the landscape, transformed the living space into a canvas where sunlight painted ever-changing masterpieces.

As Mary looked out from her window, a panoramic view of the meadow greeted her. Wildflowers swayed in the gentle breeze, and a distant creek murmured its soothing melody. The world beyond her window seemed like a living, breathing poem, and Mary felt a profound sense of peace settling within her, as though she had finally found the missing piece of her soul.

The construction of the barndominium had been an adventure, a collaboration of Mary's vision and the skilled hands of craftsmen who understood the language of wood and nails. The exposed beams inside bore witness to the strength and character that emerged from years of facing challenges head-on. The fireplace, a central fixture, became the heart of the home, radiating warmth and nostalgia, much like Mary herself.

Moving day was a celebration, a culmination of dreams taking shape. Friends and family gathered to witness the transformation of a mere structure into a haven filled with the echoes of laughter and the warmth of shared moments. Mary stood at her window, overlooking the meadow bathed in the soft hues of the setting sun, and felt a profound sense of gratitude for the community she had built.

Days turned into nights, and Mary found solace in the routine of her new life. The window became her refuge, a place to witness the ever-changing tapestry of nature. In the mornings, the sun-kissed the meadow awake, and in the evenings, a symphony of crickets serenaded the night. Each day brought a new chapter, a new opportunity to savor the simple joys that had eluded her in the hustle of city life.

With the turning of the seasons, Mary's connection to her surroundings deepened. Spring brought a burst of colors, summer painted the meadow in vibrant hues, autumn adorned the landscape in warm tones, and winter turned everything into a serene, snow-covered wonderland. Mary embraced the rhythm of nature, finding in its cycles a reflection of her journey of growth and renewal.

Mary's story echoed beyond the walls of her barndominium. Her transformation from a city dweller to a guardian of the meadow became an inspiration for those seeking a more authentic way of life. The simplicity of her existence, rooted in the appreciation of nature's beauty and the value of community, sparked a quiet revolution in the hearts of those who crossed her path.

As Mary looked out of her window, her eyes filled with gratitude, she knew that the dream barndominium was not just a house—it was a vessel for a life well-lived. In the reflection of the meadow, Mary saw not just the landscape, but a reflection of her resilience, dreams realized, and the enduring spirit of a woman who had dared to carve her own path.

Introduction

We invite you to embark on a journey into the heart of a housing trend that's sweeping across the nation – the barndominium. Born from the marriage of barn aesthetics and modern luxury, the barndominium concept offers a lifestyle that seamlessly blends the simplicity of rural living with the comfort of contemporary design.

Imagine waking up to the gentle creaking of barn doors, the scent of fresh hay lingering in the air, and the soft glow of morning sunlight streaming through expansive windows. Now, envision stepping into a space that seamlessly transitions from a rustic barn atmosphere to a sleek, modern interior. Barndominium living is an ode to the past with a nod to the future – a harmony of old-world charm and new-age functionality.

As we delve deeper into this unique dwelling, we'll explore the practicality and feasibility of turning this dream into a reality. From the initial inspiration to the construction process, we aim to guide you through the steps of creating your barndominium – a space that not only serves as a shelter but becomes an extension of your personality and aspirations.

Join us as we uncover the allure of Barndominium Living, where the rugged allure of a barn converges with the comforts of a modern home. Whether you're seeking a weekend retreat, a primary residence, or an investment opportunity, this chapter sets the stage for a journey that promises fulfillment, inspiration, and the joy of living in a space that is as unique as you are. Welcome to the world of Barndominium Living – where rustic charm meets modern amenities in perfect harmony.

In the vast landscape of home design and architecture, there exists a unique and captivating concept that has been gaining popularity in recent years – Barndominium living. A portmanteau of "barn" and "condominium," the barndominium is a distinctive type of dwelling that seamlessly blends rustic charm with modern amenities. In this chapter, we'll delve into the essence of barndominium living, exploring its origins, characteristics, and the growing movement of individuals choosing to make a barndominium their home.

Unveiling the Barndominium Aesthetic

The term "barndominium" may be relatively new to some, but the concept has deep roots in the American countryside. Traditionally, barns were structures designed for the housing of livestock, storage of crops, and protection of farming equipment. However, as times evolved and agricultural practices transformed, some visionary individuals saw potential beyond the utilitarian purpose of these barns.

The idea of converting barns into living spaces took root, blending the sturdy, weathered exteriors of barns with comfortable, modern interiors. The result? A striking fusion of the old and the new – the barndominium. The charm of weathered wood, the nostalgia of open spaces, and the allure of a simpler lifestyle all play into the unique aesthetic that defines barndominium living.

The Allure of Rustic Charm

One of the primary draws of barndominium living is the undeniable rustic charm it exudes. Imagine stepping into a space where weathered wooden beams soar overhead, where the scent of aged timber permeates the air. The very essence of the countryside is encapsulated within the walls of a barndominium, creating an environment that beckons individuals seeking a departure from the sterile and conventional.

The exterior of a barndominium often retains the barn's original features – steep gabled roofs, large sliding doors, and a facade that reflects the passage of time. This rustic allure is not merely superficial; it's a reflection of a bygone era, a nod to the simplicity and authenticity that characterize rural living.

The Modern Twist

While the exterior pays homage to the past, the interior of a barndominium seamlessly integrates modern amenities, creating a living space that is both charmingly rustic and comfortably contemporary. Open floor plans, expansive windows, and state-of-the-art appliances harmonize with the traditional elements, offering the best of both worlds.

Picture a spacious living room bathed in natural light streaming through large windows, a sleek kitchen with modern appliances nestled beneath aged wooden beams, and bedrooms that provide a cozy retreat within the embrace of weathered walls. It's this balance between the old and the new that sets barndominium living apart, making it an appealing option for those who crave a distinctive living experience.

Feasibility and Adaptability

One might wonder about the feasibility of transforming a barn into a comfortable dwelling. Surprisingly, the adaptability of barndominiums makes them an accessible and realistic option for a wide range of individuals. Unlike traditional home construction, where the process is often lengthy and resource-intensive, converting a barn into a barndominium can be a more straightforward endeavor.

The existing structure of a barn provides a solid foundation, reducing the need for extensive groundwork. This inherent stability not only expedites the construction process but also makes barndominium projects more cost-effective compared to building a new home from scratch. As a result, the dream of living in a barndominium becomes not only feasible but also financially sensible for many.

The Appeal of DIY Barndominium Projects

For those with a penchant for hands-on projects and a desire to infuse their living space with personal touches, the prospect of a DIY barndominium project is particularly alluring. The adaptability of barn structures allows for a wide range of customization options, from the layout of rooms to the choice of materials.

Embarking on a DIY barndominium project is a journey of creativity and craftsmanship. It involves repurposing and reimagining a space, infusing it with character and individuality. The sense of accomplishment that comes from transforming a humble barn into a unique, comfortable home is unparalleled and adds to the overall appeal of barndominium living.

Embracing the Barndominium Lifestyle

Beyond the physical structure, barndominium living embodies a lifestyle that embraces simplicity, connection to nature, and a sense of community. The open design of these homes fosters a flow between indoor and outdoor spaces, inviting residents to immerse themselves in the natural surroundings.

Imagine waking up to the sound of birdsong, sipping coffee on a porch that overlooks rolling fields, or stargazing on a clear night far away from the city lights. Barndominium living invites individuals to slow down, appreciate the beauty of the land, and forge a deeper connection with the environment.

Sustainable Living in Barndominiums

In an era where sustainability is at the forefront of societal concerns, barndominiums inherently offer eco-friendly advantages. The repurposing of existing structures reduces the demand for new construction materials, minimizing the environmental impact. Additionally, the spacious designs of barndominiums often incorporate energy-efficient features, such as ample natural lighting and ventilation, further contributing to sustainable living.

The move towards sustainability aligns seamlessly with the ethos of barndominium living. Embracing a lifestyle that values simplicity and connection to the land inherently fosters a sense of responsibility towards the environment. As a result, individuals choosing barndominium living often find themselves not only creating a unique home but also making a conscious choice to live more sustainably.

Barndominiums: A Growing Movement

The appeal of barndominium living has sparked a growing movement, with individuals across the country and beyond recognizing the unique blend of charm and practicality that these homes offer. From the vast expanses of rural landscapes to suburban enclaves, barndominiums have found their place as an alternative and alluring housing option.

The internet has played a significant role in fueling the barndominium movement, with online platforms showcasing stunning transformations, sharing DIY tips, and connecting like-minded individuals. Social media groups dedicated to barndominium living have become virtual communities where enthusiasts exchange ideas, seek advice, and celebrate the beauty of these distinctive homes.

The Next Chapter: Your Barndominium Journey

As we conclude this introductory chapter, it's clear that barndominium living is more than just a housing trend; it's a lifestyle choice that resonates with those seeking a harmonious blend of rustic charm and modern comfort. Whether you're captivated by the aesthetic allure, enticed by the feasibility of the construction process, or drawn to the idea of sustainable living, the world of barndominiums holds a unique promise.

In the chapters that follow, we will delve deeper into the various facets of barndominium living – from the nitty-gritty details of construction to the interior design principles that bring these spaces to life. We'll explore real-life stories of individuals who have embarked on their barndominium journey, sharing their challenges, triumphs, and the lessons learned along the way.

So, welcome to the world of Barndominiums!

Chapter 1

Site Selection and Purchase

In this chapter, you'll learn the importance of careful site selection and the factors that come into play when choosing the ideal land for your barndominium.

Factors to Consider When Buying Land

The perfect barndominium begins not with the structure itself, but with the land it's built on. A well-chosen plot can enhance your living experience, contribute to long-term property value, and even simplify the construction process.

Soil Quality

As the literal foundation of your future home, the soil quality is not something to be overlooked. Different soil types can influence the stability of your foundation, affect drainage around your property, and even dictate the types of plants you can grow.

Sandy soil, for example, is excellent for drainage, but might not be stable enough for a heavy structure. Clay soil, on the other hand, can provide stability but tends to retain water, which could lead to foundation issues down the line. It's crucial to get a soil test done before making your purchase to ensure the land is suitable for building.

Accessibility

Your dream barndominium might be perched on a hillside with panoramic views or nestled in a secluded woodland, but consider how easy it will be to access. Think about the convenience of everyday commuting, the availability of road infrastructure, and the potential costs of bringing in construction materials.

Moreover, think about accessibility in terms of seasons. A dirt road might be easy to navigate in summer, but what about in heavy rain or snow?

Proximity to Utilities

Off-grid living is an appealing concept, but the reality of being without power, water, or the internet can be quite challenging. When choosing your land, consider the proximity to utilities.

Bringing in utilities to a remote location can be costly and complicated, involving digging trenches, laying cables, and potentially negotiating with neighbors for easements. You might find a beautiful piece of land at a bargain price, only to discover the cost to connect utilities is prohibitive. We will revisit utilities in a later chapter.

Local Climate

The local climate plays a significant role in your living experience and the design of your barndominium. It affects your heating and cooling needs, the type of insulation you'll need, and even the materials you should use for construction.

For instance, if you're building in a region with harsh winters, you'll want to consider high-quality insulation and a heating system that can keep up with the cold. On the other hand, in a hot, dry climate, you might prioritize shade, ventilation, and a cool roofing material.

Surrounding Environment

Lastly, take a look at the surrounding environment. Is the land near a flood zone or a wildfire-prone area? Are there potential sources of noise or pollution nearby? How about the view - will you be looking out at a serene forest or a busy highway?

The surrounding environment not only affects your living experience but can also have a significant impact on the value of your property. A tranquil setting with a beautiful view can make your barndominium a highly sought-after property should you ever decide to sell.

Choosing the right land for your barndominium is a crucial first step in the building process. By considering these factors, you'll be well-equipped to find a plot that suits your needs, meets your budget, and provides a solid foundation for your dream home.

Understanding Land Surveys

Once you've zeroed in on an ideal piece of land, the next crucial step is to get a thorough land survey done. A land survey offers an intricate look at the property, providing essential details that significantly influence the planning and execution of your barndominium project.

Boundary Lines

The first piece of information a land survey provides is the exact boundaries of the property. Knowing the precise perimeters of your land is fundamental. It affects everything from the placement of your barndominium to the design of outdoor spaces like gardens and patios.

Understanding boundary lines also helps avoid potential disputes with neighbors in the future. Even if the current owners assure you of the boundaries, it's wise to get a land survey done, as sometimes, physical markers like fences or hedgerows may not accurately represent the legal boundaries.

Easements and Rights of Way

Another critical aspect that a land survey reveals is the presence of any easements or rights of way. An easement is a legal right for someone else to use a part of your property for a specific purpose, like a shared driveway or a public footpath.

Knowing about these beforehand is crucial, as it may affect your building plans and privacy. For instance, if there's a utility easement across your property, you won't be able to build on that particular section. Similarly, a right of way for a neighbor could influence the privacy of your home.

Potential Building Sites

A detailed land survey can also help identify potential building sites on your property. The surveyor can mark out flat areas, slopes, and other physical features that could influence where you build your barndominium.

For instance, a slight rise could provide impressive views from your living room. On the other hand, a flat area might be ideal for an outdoor patio or garden. Thus, a land survey can be a valuable tool in the initial design and planning stages of your project.

Natural Features

Finally, a land survey provides a detailed record of the natural features of the land, like trees, streams, rocks, and the contour of the terrain. This information is critical not just for planning the building's location, but also for designing the overall layout of your property.

For example, a beautiful old tree could become a focal point in your garden or a naturally occurring slope could be used for a walkout basement. Incorporating these natural features into your design can enhance the charm and uniqueness of your barndominium.

Getting a detailed land survey is a worthwhile investment that can save you potential legal troubles, help you make informed design decisions, and ensure the smooth execution of your building project. With this foundation, we will move forward to the art of negotiation in land purchasing.

Negotiating a Land Purchase

Now that you've identified the perfect piece of land and have a thorough understanding of its features and boundaries, the next stage is to negotiate the purchase. This process involves several steps, each of which plays a critical role in ensuring you get the best deal possible.

Market Research

Before you start negotiating, it's crucial to have a clear understanding of the current real estate market in the area where your prospective land is located. This involves researching recent land sales in the vicinity to get an idea of pricing trends.

You might also want to look into future development plans for the area as these can significantly impact the value of your land. If there are plans for infrastructure development like new roads or utilities, this might increase the value of your property. Conversely, if a large commercial development is planned nearby, this might negatively impact the tranquility of your future barndominium.

Property Appraisal

Once you have a fair idea of the market value, the next step is to have the property appraised. An appraisal is a professional assessment of the property's value based on various factors like size, location, accessibility, and the presence of any structures or utilities.

While the seller might provide you with an appraisal, it's often a good idea to get an independent appraisal. This can help you ensure that the asking price is fair and provide a solid basis for negotiation.

Making an Offer

With your market research and appraisal in hand, you're now ready to make an offer. This is where your negotiation skills come into play.

Remember, the asking price is just that - what the seller is asking for. It's not necessarily what they expect to get, nor what the property is worth. Don't be afraid to offer less than the asking price if your research and appraisal suggest that's fair.

When making an offer, it's also a good idea to outline any conditions you have. For example, you might make your offer contingent on obtaining a satisfactory soil test or securing financing.

Closing the Deal

Once your offer is accepted, the final step is to close the deal. This involves signing a contract of sale, paying a deposit, and arranging for the payment of the balance.

It's also when any conditions you've stipulated, such as securing financing or conducting further inspections, need to be fulfilled. If any legal issues need to be resolved, such as easements or rights of way, this is when that happens.

Once the balance of the purchase price is paid, the property deed will be transferred to your name. Congratulations, you're now the proud owner of land for your barndominium!

While these steps might seem daunting, remember that each one is a step closer to realizing your dream of barndominium living. By doing your homework, staying informed, and negotiating effectively, you can secure the perfect piece of land at a fair price. Now, it's time to roll up your sleeves and get ready for the next stage - preparing your land for construction.

Preparing Your Land for Construction

With the land purchase complete, it's time to start preparing your property for the construction of your dream barndominium. This stage involves a few critical steps that ensure your land is ready for the building process.

Clearing and Grading

Land clearing is the first step in preparing your land for construction. This process involves removing trees, shrubs, rocks, and other obstacles that may hinder construction. While this may seem like a straightforward task, it requires careful planning to ensure minimal environmental impact and to preserve any natural features you'd like to incorporate into your landscape design.

After clearing the land, the next step is grading. This involves leveling the land and shaping it to suit the design of your barndominium and the surrounding landscape. Proper grading is vital to ensure adequate drainage around your property and to prevent potential issues like water pooling or soil erosion. Professional land grading can also help create beautiful, functional outdoor spaces for gardens, patios, and driveways.

Soil Testing

You've already conducted a basic soil test during the land purchase process. However, now that you're moving into the construction phase, a more detailed soil test is in order. This test will assess the soil's capacity to support your barndominium's weight, its drainage properties, and its potential for expansion and contraction, which could impact the stability of your foundation.

Engage a professional geotechnical engineer to conduct this test. They will take soil samples from different parts of your property and at various depths to provide a comprehensive understanding of the soil conditions on your land.

Establishing Access Roads

Next, consider how construction materials and equipment will reach your building site. If your plot is in a remote location or surrounded by rough terrain, you may need to build access roads.

The design of these roads will depend on the topography of your land, the type of vehicles that will be using them, and the local climate. For instance, a steep slope may require a winding road to reduce the gradient, while areas with heavy rainfall might need a well-drained gravel road to prevent it from becoming muddy and impassable.

Setting Up Temporary Utilities

Lastly, setting up temporary utilities is an essential part of preparing your land for construction. These might include a temporary water supply for construction needs, electrical hookups for power tools, and portable toilets for the construction crew.

You may also need to install a temporary construction fence for safety and security purposes. This fence can help keep children and pets out of the construction site, deter thieves from stealing construction materials, and prevent unauthorized individuals from accessing the site.

Bringing your dream barndominium to life involves much more than just building the structure itself. It's about creating a home that's perfectly suited to its surroundings and designed to enhance your lifestyle. By carefully preparing your land for construction, you're not just laying the groundwork for your barndominium - you're setting the stage for a home that's truly a part of the landscape. With your land cleared, graded, and ready for construction, the vision of your barndominium is one step closer to becoming a reality.

Chapter 2

Designing Your Barndominium

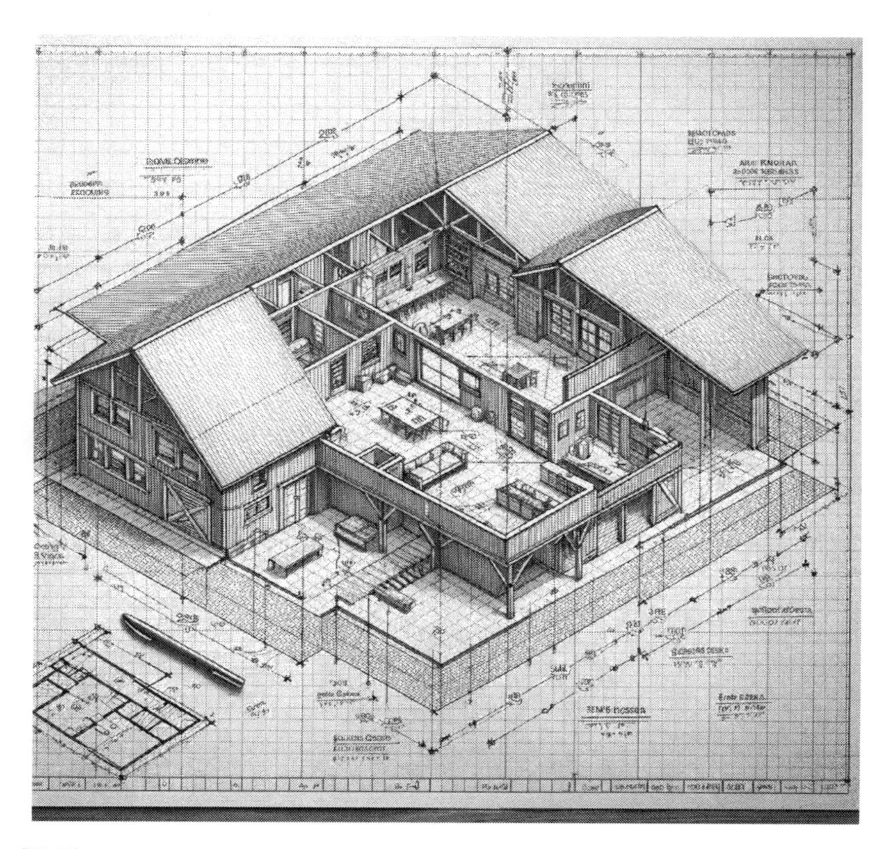

The blank canvas of your new barndominium waits, ready to be transformed into the home of your dreams. But this isn't about creating a carbon copy of someone else's design - this is about tailoring every nook and cranny to your unique lifestyle needs. Many Barndo designs start simple. Don't be afraid to draw out your thoughts on whatever you have handy and near you. Inspiration can come at any time. From a napkin to a piece of graph paper, when an idea comes, write it down. A majority of layout designs have had numerous amendments before they get a final stamp. Let's start the process of molding this canvas to fit your life like a glove.

Post Frame or Stick Built

As we delve into the realm of construction, understanding the fundamental differences between post-frame buildings and traditionally stick-built structures is pivotal. These two methods represent distinct approaches to creating durable and functional spaces. Let's explore the characteristics that set them apart and the considerations that influence the choice between these construction techniques.

Structural Design: Simplicity vs. Complexity

Post Frame Construction:

Post-frame structures are known for their simplicity. Vertical posts, typically set deep into the ground and anchored to a foundation, serve as the primary load-bearing elements. These posts are spaced at intervals to support horizontal framing members, forming a skeleton that carries the building's weight. The open interior allows for versatile layouts without the need for interior load-bearing walls.

Stick Built Construction:

Traditionally stick-built structures rely on a more intricate framework. This method involves constructing a frame using dimensional lumber or engineered wood products, with vertical studs forming the primary load-bearing members. The frame consists of multiple interconnected components, including studs, joists, and rafters, requiring precision and attention to detail.

Foundation Requirements: Versatility vs. Rigidity

Post Frame Construction:

Post-frame buildings are renowned for their flexibility in foundation options. They can be constructed on various foundation types, including concrete slabs, piers, or even directly on the ground. The load distribution through vertical posts enables adaptability to different soil conditions, making post-frame structures suitable for a range of terrains.

Stick Built Construction:

Stick-built structures typically demand a more rigid foundation, often requiring deep footings or concrete slabs. The intricate framework necessitates a foundation that can evenly distribute the load across the structure's footprint. This requirement limits the adaptability of stick-built structures to certain types of foundations.

Construction Speed: Efficiency vs. Precision

Post Frame Construction:

One of the significant advantages of post-frame construction is its efficiency. The straightforward assembly of vertical posts and horizontal framing members allows for rapid construction. The larger structural components mean fewer pieces to assemble, reducing construction time significantly.

Stick Built Construction:

Stick-built structures, with their intricate framing and numerous smaller components, often require more time for construction. Precision in cutting and placing each piece is crucial, and the process can be more time-consuming compared to the simplicity of post-frame construction.

Cost Considerations: Affordability vs. Customization

Post Frame Construction:

Post-frame buildings are often more cost-effective due to their efficient use of materials and quicker construction time. The larger spacing between posts and fewer load-bearing elements contribute to reduced material costs. Additionally, the simplified construction process minimizes labor expenses.

Stick Built Construction:

While stick-built structures offer greater customization options, this often comes at a higher cost. The precision required in constructing the intricate frame, the need for a more rigid foundation, and longer construction times contribute to increased material and labor expenses.

Interior Space: Openness vs. Partitioning

Post Frame Construction:

The open nature of post-frame construction allows for expansive, unobstructed interior spaces. With no need for interior load-bearing walls, the layout can be easily adapted to various uses, making post-frame structures ideal for large storage areas, workshops, or agricultural buildings.

Stick Built Construction:

Stick-built structures often require interior load-bearing walls to support the complex framework. This can limit the flexibility of interior layouts, requiring careful planning and design to accommodate specific needs.

In the dynamic world of construction, choosing between post-frame and stick-built structures hinges on factors such as design preferences, budget considerations, and the intended use of the space. Post-frame construction offers simplicity, cost-effectiveness, and rapid assembly, making it an excellent choice for various applications. On the other hand, stick-built structures provide greater customization options and meticulous craftsmanship, catering to those who prioritize intricate designs and architectural flexibility. Understanding the nuances of each method empowers builders and property owners to make informed decisions that align with their vision and practical requirements.

Understanding Your Lifestyle Needs

The magic of a barndominium lies in its ability to be anything you want it to be. From a bustling family hub to a tranquil artist's retreat, your barndominium should be an extension of your lifestyle. Here's how to start defining what you need.

Space Requirements

First, consider the size of your home. Do you envision a cozy and intimate space or a sprawling open-plan layout? The number of bedrooms, bathrooms, and the size of communal areas should depend on your lifestyle and the size of your family.

Think about your daily routines and how you use space. If you love cooking, a spacious kitchen might be a priority. If you have children, a large, safe outdoor play area might be crucial. If you're a fitness enthusiast, space for a home gym could be on the cards. Consider where you are going to spend most of your time while at home.

Take your time in considering room sizes. Once your walls start going up, it becomes much more difficult to add a walk-in closet or to make a master bathroom larger. Make certain that all your needs fit within the footprint of your build, and allow enough square footage to be sure that you are building your dream Barndo.

The size of your shop (garage) is an extremely important consideration while planning your space requirements. For some, this space will need to be larger than for others. If you are a family with lots of toys (side-by-sides, 4 wheelers, tractors, motorcycles, boats, RV's) you might require a substantially larger shop size. Some design ideas even include a separate building, possibly connected with an annex or breezeway, for a large shop.

Future Family Growth

A home isn't just for today, it's for tomorrow as well. If you're planning to have children or expecting your family to grow in the future, your design should be able to accommodate these changes. Consider incorporating extra bedrooms, or designing spaces that can be easily converted in the future, like a home office that could become a nursery.

Work-from-Home Needs

The trend of remote work is on the rise, and a dedicated workspace is a significant factor to consider. If you or your spouse work from home, think about incorporating a quiet, well-lit home office into your design. Consider the location of this space; it should be far enough from the noise and bustle of the household but close enough for you to feel connected.

Outdoor Living Preferences

Your barndominium's design isn't limited to its four walls. The outdoor spaces are just as important. Consider your climate and how much time you could comfortably spend outside. Do you envision summer barbecues on a patio or cozy winter nights around a fire pit? Would a vegetable garden or chicken coop be part of your sustainable living plan?

Understanding Your Lifestyle Needs: A Checklist

To help you define your lifestyle needs, here's a checklist to get you started:

- Determine the total size of your home, the number of bedrooms and bathrooms, and the size of communal areas.
- Plan for future family growth by incorporating extra bedrooms or convertible spaces.
- If you work from home, ensure you include a quiet, well-lit home office in your design.
- Consider your outdoor living preferences and how they can be incorporated into your design.

In the end, your barndominium should be a reflection of who you are and how you live. By taking the time to understand your lifestyle needs, you can ensure your new home will serve you faithfully, not just for today, but for many years to come.

Creating Functional Layouts

The heart of a barndominium beats in its layout. The arrangement of your rooms and spaces plays a crucial role in the functionality and comfort of your home. With this in mind, let's explore how to create a layout that marries form and function.

Open Floor Plan Concepts

Pioneered by Frank Lloyd Wright, open floor plans break down the barriers between traditional rooms to create a sense of flow and connectivity. This concept is a natural fit for barndominiums, given their origins in spacious barn structures.

An open floor plan can make your home feel larger and brighter by allowing natural light to filter through the entire space. It also fosters a sense of togetherness, making it easier to interact with family members even when you're in different areas of the home.

However, designing an effective open floor plan requires careful thought. You'll need to define distinct areas for different activities, such as cooking, dining, and relaxing, while maintaining the overall sense of openness. This can be achieved through the strategic placement of furniture, the use of area rugs, or subtle changes in ceiling height or flooring materials.

Private vs. Public Spaces

While open floor plans are great for fostering connectivity, there's also a need for private spaces in your home. Bedrooms, bathrooms, and home offices typically require more privacy and should be located away from the main communal areas.

One effective strategy is to group public and private spaces into separate zones. For instance, you could locate all the bedrooms at one end of your barndominium and the living, dining, and kitchen areas at the other. This not only enhances privacy but also helps to reduce noise transfer between these zones.

Storage Solutions

One often overlooked aspect of home design is storage. A well-designed home should have ample storage to keep your living spaces clutter-free and functional.

In a barndominium, you have the opportunity to get creative with your storage solutions. You could incorporate built-in shelves in your walls, use the space under your stairs, or even design a loft area for extra storage.

Don't forget to consider the specific storage needs of each room. The kitchen will require space for appliances and groceries, bedrooms need closets for clothes, and a home office will need storage for documents and supplies.

Indoor-Outdoor Flow

Given the rural setting of most barndominiums, it's worth considering how your home can connect with its surroundings. A strong indoor-outdoor flow can enhance the sense of space in your home and allow you to make the most of your outdoor areas.

Large windows and doors can provide views of the landscape and allow natural light to flood your living spaces. Outdoor living areas, such as patios or decks, should be easily accessible from the main living areas.

You could also consider incorporating natural materials, such as wood or stone, in your interior design to further blur the line between indoors and outdoors.

Designing your barndominium layout is a bit like choreographing a dance. Each element must move in harmony with the others, resulting in a home that is not only beautiful but also perfectly attuned to your lifestyle. By considering these principles, you can create a functional layout that brings your vision of barndominium living to life.

Choosing the Right Building Materials

Deciding on the appropriate materials for your barndominium is akin to selecting the ingredients for a special meal. Each one plays a pivotal role in the outcome, influencing the strength, cost, beauty, and ecological footprint of your home. Let's explore these factors in greater detail.

Durability

The longevity of your barndominium is directly related to the durability of the construction materials. Durable materials resist wear and tear, withstand extreme weather conditions, and require less maintenance, which can save you time and money in the long run.

In the skeleton of your barndominium, for example, steel beams provide a high strength-to-weight ratio, making them an excellent choice for the structure. For the exterior, metal sidings can resist the harshest elements, while their fire-resistant properties add an extra layer of safety.

Inside your home, materials also need to stand up to daily use. For instance, hardwood flooring or ceramic tiles can be a wise choice for high-traffic areas, offering both resilience and easy maintenance.

Cost-Effectiveness

While building your dream barndominium, it's important to strike a balance between quality and affordability. The most expensive materials aren't always the best fit for your needs, and many cost-effective alternatives provide excellent performance and aesthetics.

For example, in your roofing, traditional materials like slate or cedar can be costly and require professional installation. On the other hand, metal roofing is more affordable, easier to install, and offers comparable durability and aesthetic appeal.

Similarly, for interior walls, drywall is a cost-effective option that provides a smooth surface for paint or wallpaper and offers good fire resistance. For insulation, spray foam, while a bit more expensive upfront, provides high thermal resistance and can result in long-term energy savings.

Aesthetic Appeal

Your barndominium should be a feast for the eyes, a place that makes your heart sing every time you come home. The materials you choose play a significant role in determining the look and feel of your home.

For instance, a reclaimed wood accent wall can add warmth and rustic charm to your living room, while granite countertops in the kitchen exude elegance and timeless appeal.

Consider the visual impact of each material and how it contributes to the overall style you're aiming for. Remember, beauty is in the details. A well-chosen light fixture, a stylish cabinet knob, or a unique tile design can make a world of difference in your home's aesthetic appeal.

Environmental Impact

Building your barndominium is not just about creating a home for you, but also about caring for the home we all share - our planet. The materials you choose can have a significant impact on the environment, both during the construction process and throughout your home's lifespan.

Choosing locally sourced materials can reduce the carbon footprint associated with transportation. Using recycled or reclaimed materials can help reduce waste and save valuable resources. For instance, reclaimed wood can be used for flooring, beams, or furniture, adding character to your home while being eco-friendly.

Consider also the energy efficiency of materials. Insulation materials with high R-values reduce heat transfer, helping to lower energy consumption for heating and cooling. Double-glazed windows can help keep your home comfortable while reducing energy use.

In essence, the materials you choose for your barndominium are the building blocks of your future home. They determine not only the strength and cost of your home but also its beauty and impact on the environment. By carefully considering each of these factors, you can make informed decisions that align with your values, meet your needs, and bring you one step closer to your dream barndominium.

Planning for Future Expansion

Building a barndominium is an exciting venture, a unique opportunity to craft a home that perfectly aligns with your current lifestyle. But life is fluid and ever-changing. What suits you now may not be ideal in a few years. That's why it's crucial to incorporate flexibility in your design, allowing your barndominium to evolve as your life does.

Flexible Design Elements

One way to future-proof your barndominium is to integrate flexible design elements. These are features that can be easily modified or repurposed to meet changing needs.

For example, consider a large open space that could serve as a living room now, but could easily be partitioned into bedrooms or a home office later. Or a spacious loft area that could be a storage space now but transformed into a cozy guest room in the future.

A clever approach is to include movable walls in your design. These allow you to easily reconfigure your space as needed. For instance, you could use a movable wall to enlarge a guest room when a family comes to visit, then shift it back to create a larger living area when they leave.

Another flexible design element is the use of modular furniture. This type of furniture can be rearranged, expanded, or downsized to adapt to your changing needs. It's not only versatile but also a great way to maximize space in your barndominium.

Pre-Planned Additions

Another strategy for planning for future expansion is to incorporate pre-planned additions into your design. These are areas of your home that are designed to accommodate future expansions easily.

A common example is a 'bonus room' over the garage. This space could be left unfinished for now, keeping initial construction costs down, but designed in such a way that it could be easily finished and converted into a bedroom, home office, or playroom in the future.

Another approach is to design your barndominium with the potential for physical extensions. For instance, you could plan for doors or windows to be replaced with connecting hallways to future rooms. Or design your landscaping to accommodate a future patio or deck.

Potential for Upgrades

Lastly, consider the potential for upgrades. As technology advances and your budget allows, you might want to add high-end features to your barndominium.

For instance, you might start with standard appliances in your kitchen, with the plan to upgrade to professional-grade ones in the future. Or perhaps you'll want to add a luxury feature like a hot tub or sauna down the line.

Incorporating conduits into your walls during construction can make future electrical upgrades much easier and less invasive. Similarly, making sure your current HVAC system has the capacity to handle potential future additions can save you a major headache later.

In conclusion, designing your barndominium with the future in mind requires a blend of creativity, foresight, and practicality. By incorporating flexible design elements, planning for additions, and considering potential upgrades, you can ensure your barndominium remains a perfect fit for your lifestyle, not just for now, but for many years to come.

As we wrap up this chapter, remember that every design decision you make is another stitch in the tapestry of your future life, a step towards a home that truly reflects who you are. In the next chapter, we'll shift our focus from the design to the practical aspects of bringing your dream to life, navigating permits, and regulations.

Chapter 3

Navigating the Maze of Permits and Regulations

Beware, aspiring barndominium builders! There lurks a labyrinthine maze in your path to rural bliss. It's a necessary hurdle, filled with building codes, zoning laws, permits, and inspections. Online groups and chat communities are filled with examples where bardno's are not zoned and permitted correctly.

Or where a lender determines that a post frame home is not eligible for a construction loan. Some barn home buildings never get further than an idea because of the fear that the project will not be legally considered a residential dwelling. Fear not, for this chapter will serve as your trusted compass, guiding you through the tangled web of red tape, ensuring you emerge victorious on the other side.

While it may seem daunting, mastering the art of permits and regulations is integral to your barndominium project. These laws ensure your home is safe, environmentally friendly, and in harmony with its surroundings. Let's roll up our sleeves, pull out our compass, and prepare to conquer this maze, one step at a time.

Understanding Building Codes and Zoning Laws

One of the first steps on your journey is understanding the ins and outs of building codes and zoning laws. These rules, set by local governments, dictate what you can build, where you can build it, and how it must be built.

Residential vs. Agricultural Zoning

Zoning laws divide land into zones, each with its own set of rules. The two most common zones you'll encounter when building a barndominium are residential and agricultural.

Residential zoning typically allows for single-family homes, duplexes, apartments, and other dwellings. Agricultural zoning, on the other hand, is usually reserved for farming activities. However, many areas allow for residential dwellings on agriculturally zoned land, often with more relaxed rules regarding building size and placement.

Therefore, when buying land for your barndominium, it's crucial to check its zoning classification. Remember, the zoning not only affects whether you can build your barndominium, but also potentially impacts things like property taxes, the keeping of livestock, and future resale value.

Building Height Restrictions

Building codes often impose limits on building height. These restrictions are designed to maintain a consistent scale within neighborhoods, protect views, and ensure the safety of structures.

Imagine building your dream barndominium, only to find out that your majestic, vaulted ceilings exceed local height restrictions. To avoid this pitfall, check with your local planning department for any height restrictions before finalizing your design.

Setback Requirements

Setbacks refer to the distance a building must be set back from property lines, roads, or other features. These requirements help prevent overcrowding, protect natural features, and maintain privacy between neighbors.

For example, your local code might require your barndominium to be set back at least 20 feet from the front property line, 10 feet from the side property lines, and 25 feet from any water bodies. It's important to familiarize yourself with these requirements early on, as they can significantly influence where you can place your building on your land.

Environmental Regulations

Lastly, environmental regulations play an increasingly important role in building codes. These rules aim to protect sensitive natural areas, promote energy efficiency, and ensure healthy indoor environments.

For instance, if your property contains wetlands, there could be restrictions on building near these areas. Similarly, there might be requirements for insulation levels, water efficiency, and the use of low-VOC materials in your barndominium.

Properly understanding and navigating building codes and zoning laws is akin to decoding a complex puzzle. It requires patience, attention to detail, and a good dose of perseverance. However, cracking this code is a crucial step on your path to building a barndominium that is safe, legal, and harmonious with its surroundings.

As we delve deeper into this maze, keep in mind the importance of every rule and regulation. They aren't just arbitrary obstacles in your path. Instead, they are carefully designed safeguards, ensuring that your dream barndominium becomes a beloved part of the landscape, standing strong and beautiful for generations to come.

Applying for Construction Permits

Navigating the sea of paperwork for construction permits might feel like an uphill battle, but it is an integral part of the barndominium building process. With careful preparation and a touch of patience, this challenge becomes a surmountable task. Let's break it down into manageable steps.

Preparing Application Documents

Application documents are your passport to the world of construction. They offer a detailed snapshot of what you plan to build, how you plan to build it, and how it complies with local regulations.

A complete application usually includes your site plan, floor plan, and elevations, along with a detailed description of the materials and construction methods you intend to use. It's a good idea to involve your architect or builder in preparing these documents, as their expertise can help ensure accuracy and compliance with local codes.

Navigating the Review Process

Once you've submitted your application, you enter the review process. During this phase, your local planning department will scrutinize your plans, checking for compliance with building codes, zoning regulations, and environmental guidelines.

This review process can take several weeks or even months, depending on the complexity of your project and the workload of the planning department. Patience certainly becomes a virtue, but it's also essential to be proactive. Regularly follow up with the planning department to check the status of your application and respond promptly to any inquiries they might have.

Addressing Feedback and Revisions

Feedback is an inevitable part of the review process. You might receive requests for additional information, suggestions for changes, or even objections to certain aspects of your design.

While it can be disappointing to receive negative feedback or requests for revisions, it's essential to approach them with an open mind. Remember, the goal of the planning department is not to thwart your dreams but to ensure your barndominium is safe, sustainable, and harmonious with its surroundings.

When addressing feedback, work closely with your architect or builder. They can help you understand the implications of the requested changes and come up with solutions that satisfy the planning department while keeping your vision intact.

Securing Final Approval

With your revisions submitted and all concerns addressed, you're on the home stretch. The planning department will conduct a final review of your application. If all is in order, they'll issue a building permit, giving you the green light to start construction.

Securing final approval is a moment of triumph, a significant milestone on your barndominium building adventure. It's a testament to your diligence, patience, and commitment to creating a home that meets not only your dreams but also the stringent standards of safety and sustainability.

The process of applying for construction permits might seem like a monumental task, but it's simply a series of steps, each leading you closer to your goal. With careful preparation, a proactive approach, and a willingness to adapt, you can navigate this process successfully and secure the necessary permits to bring your dream barndominium to life.

Handling Inspections

Once construction begins on your dream barndominium, another critical element of the process comes into play - inspections. These are a series of checks conducted at various stages of the project to ensure the work complies with the approved plans and local building codes. While the inspection process may seem intimidating, it's an opportunity to ensure your barndominium is built to the highest standards of safety and quality.

Preparing for Each Inspection Phase

Inspections occur at multiple phases throughout the construction process. Each one focuses on a specific aspect of the build, such as the foundation, framing, plumbing, electrical, and finish work.

The foundation inspection, for example, checks the size and depth of the footings and the placement of the rebar. The framing inspection, on the other hand, evaluates the structural integrity of the walls, floors, and roof structure.

To ensure a smooth inspection process, you'll need to be well-prepared. This means understanding what the inspector will be looking for at each phase and ensuring the work is complete and ready for review. Keep a copy of the approved plans on-site for reference, and make sure all work is visible and accessible to the inspector.

Addressing Inspector Feedback

During each inspection, the building inspector will provide feedback. This could range from a simple approval, allowing you to proceed to the next phase, to a list of items that need correction.

If your project doesn't pass an inspection, don't panic. This is a relatively common occurrence and not a significant setback. The inspector will provide a list of issues that need to be addressed, which could include things like adding more nails to a joist hanger, installing additional bracing for a wall, or correcting a plumbing connection.

Once you've addressed the issues, you can request a re-inspection. The inspector will review the corrections and, if satisfactory, will approve the work. While addressing feedback can sometimes be time-consuming, it's a crucial part of ensuring your barndominium is built to the highest standards.

Final Inspection and Certificate of Occupancy

The final inspection is a significant milestone in your barndominium build. This inspection is a thorough review of all aspects of the project, ensuring everything has been completed correctly and safely.

The final inspection covers a broad range of items, including structural elements, electrical and plumbing systems, HVAC, insulation, and final finishes. The inspector will also check that all previous inspection issues have been correctly addressed.

If your project passes the final inspection, the building department will issue a Certificate of Occupancy. This document certifies that the building is safe to occupy and has been constructed in compliance with the approved plans and all applicable building codes.

Receiving the Certificate of Occupancy is a moment of triumph, the culmination of all the hard work, planning, and diligence that's gone into your build. It's the final official stamp of approval, allowing you to move into your dream barndominium and start the next chapter of your journey in your new home.

Navigating the inspection process is like traversing a winding river. With each bend, you uncover a new challenge, a new opportunity to learn and grow. By understanding the inspection process, being well-prepared, and addressing feedback promptly, you can ensure your barndominium is built to the highest standards, providing a safe, durable, and quality home for you and your family. So, let's strap on our life jackets, grab our oars, and get ready to navigate the riveting river of inspections!

Keeping Your Project Legal

The foundation has been poured, the beams are rising, and your dream barndominium begins to take shape. Amid the bustle of construction, it's easy to lose sight of the less tangible aspects of the process. However, staying on the right side of the law is as crucial as any physical construction activity. Let's explore how to ensure your project remains legal and respectful of community norms.

Regularly Reviewing Permit Status

Ensuring your build stays within the bounds of the law starts with keeping a close eye on your permit status. While your construction permit allows you to build your barndominium according to your approved plans, it doesn't give you carte blanche to make changes as you see fit.

If your plans evolve or you decide to add features not included in your initial application, you'll need to apply for a permit amendment. This ensures that all work done on your property remains compliant with local codes and regulations. Regular checks on your permit status, especially when changes occur, help you avoid potential legal issues down the line.

Maintaining Good Relations with Neighbors

While your barndominium might sit on your private property, it's also part of a broader community. Maintaining good relations with your neighbors is not only a matter of courtesy but can also help keep your project running smoothly.

Be forthright about your construction plans, keeping neighbors informed about project timelines, particularly for noisy or disruptive work. Minimize construction noise during early morning or late evening hours, and keep your construction site tidy to prevent any undue inconvenience.

Remember, a friendly approach goes a long way. A neighbor who feels respected and considered is less likely to raise objections or complaints about your project.

Addressing Complaints Promptly

Despite your best efforts, complaints might still arise during your construction process. It could be a neighbor upset about dust from your site, or a city official who's received a report about work being done without the proper permits.

Addressing these complaints promptly and professionally is key to keeping your project on track. If a neighbor complains, listen to their concerns and do your best to mitigate the issue. If the complaint comes from a city official, cooperate fully and take the necessary steps to rectify the situation.

Ensuring All Work Meets Code Standards

Lastly, ensuring all work done on your barndominium meets local building code standards is a non-negotiable aspect of keeping your project legal. These codes are in place to ensure the safety, health, and overall welfare of the residents.

This goes beyond just the structural integrity of your barndominium. It also includes aspects like electrical wiring, plumbing, insulation, and ventilation. Hiring experienced, licensed contractors and conducting regular inspections can help ensure all work meets the required standards.

Navigating the legal aspects of your barndominium build might not be the most glamorous part of the process, but it's a critical component of a successful project. By staying on top of your permit status, maintaining good community relations, addressing complaints promptly, and ensuring your work meets code standards, you're not just building a home - you're also building a respectful and lawful presence in your community.

So, raise your hammer high, and strike with the confidence that every nail driven, every beam raised is not just creating a structure, but a home that respects and cares for the community it stands within. As we move forward to the next chapter, we transition from the legalities and paper trails to the physicality of your dream barndominium - laying the foundation, erecting the structure, and bringing your vision to life.

Chapter 4

Building the Shell of Your Barndominium

Imagine the exhilaration of seeing your barndominium rise from the ground, each beam and wall a testament to your vision and hard work. This chapter delves into the crucial first stages of this process - laying the foundation. This is the bedrock upon which your dream home will stand, a symbol of solidity, strength, and permanence. So, let's roll up our sleeves and get down to the nitty-gritty of laying the foundation.

Laying the Foundation

Traditional Stick Built Construction:

Site Preparation

The first step in laying the foundation is preparing the site. This involves clearing any vegetation, rocks, or debris from the area where your barndominium will stand. You'll also need to level the ground to ensure a flat, stable surface for the foundation.

Picture a gardener preparing a bed for planting, removing weeds and rocks, and raking the soil to a smooth, even texture. Your job in site preparation is similar but on a larger scale.

Concrete Mixing

Once the site is ready, it's time to mix the concrete for your foundation. This is a critical step that requires precision and care.

Think of it like baking a cake. Just as you carefully measure each ingredient to ensure the perfect taste and texture, you'll need to accurately measure the cement, sand, gravel, and water for your concrete mix.

The standard ratio is 1 part cement, 2 parts sand, and 3 parts gravel, with enough water to make a smooth, pourable mixture. Mixing can be done in a wheelbarrow for small batches, but for larger volumes, consider renting a concrete mixer.

Formwork Installation

The next step is formwork installation. This involves building a temporary wooden structure to contain the concrete while it sets.

Imagine building a sandcastle on the beach. You use a bucket to hold the wet sand in place until it dries into the shape of a castle.

The formwork serves a similar purpose for your concrete foundation.

The formwork should be sturdy and well-braced to withstand the weight of the concrete. Also, consider adding a plastic sheet or other waterproofing layer inside the formwork to prevent the concrete from sticking to the wood.

Pouring and Curing

With the formwork in place, it's time to pour the concrete. Start at one corner and work your way across the form, spreading the concrete evenly as you go. Once the form is filled, use a shovel or trowel to smooth the surface of the concrete.

Pouring concrete is a bit like spreading batter in a cake pan. You want to ensure an even distribution to get a flat, level surface.

Once the concrete is poured, the curing process begins. Curing involves keeping the concrete damp for several days to allow it to harden and gain strength. You can do this by regularly spraying it with water or covering it with a curing blanket.

Laying the foundation for your barndominium is similar to setting the stage for a play. It's the platform on which the drama of construction will unfold, supporting the weight of your dream home and anchoring it firmly to the ground. With this solid foundation, your vision is one step closer to becoming a reality, ready to rise majestically from the earth.

Post Frame Pole Barn Construction:

The foundation pieces for a post frame building are the posts themselves. If your design decision is to build a post-frame style structure, the first step is to lay out where your posts (poles) will be located. A Typical span between poles is 8 feet. Always make sure to account for where overhead doors, entry doors, and windows are going to be located when laying out your poles.

Installing pole barn posts can be approached using different techniques, and the choice often depends on factors such as soil conditions, local building codes, and personal preferences. Here are some common techniques for installing pole barn posts:

Direct Embedment:

- This technique involves placing the posts directly into the ground without concrete footings.
- The depth of the hole is typically below the frost line to prevent heaving.
- Backfilling with soil provides stability, and the natural soil acts as the foundation.

Concrete Footings:

- Dig holes for the posts and pour concrete footings at the bottom of each hole.
- Place the posts on top of the cured footings.
- This technique provides additional stability and is suitable for areas with varying soil conditions.

Concrete Piers:

- Similar to concrete footings, concrete piers involve pouring a cylindrical concrete structure to support each post.
- The posts are set on top of the piers once the concrete has cured.
- This technique is effective in areas with challenging soil conditions.

Post Anchors or Brackets:

- Use metal post anchors or brackets to secure the posts to a concrete footing or pier.
- The post is bolted or otherwise fastened to the anchor, providing a connection between the post and the foundation.
- This method ensures stability and helps prevent settling.

Treated Wood or Synthetic Poles:

- In some cases, pressure-treated wood or synthetic poles designed for ground contact may be used.
- These materials are resistant to decay and insects, reducing the need for concrete footings.

Auger Anchors:

- Auger anchors are screw-like devices that are screwed into the ground beneath each post.
- These provide additional stability, especially in areas with softer soil.

Sonotube or Cardboard Forms:

- Sonotube or cardboard forms can be used to create cylindrical shapes for concrete footings or piers.
- They simplify the pouring process and help maintain a consistent shape for the foundation.

Post Drivers:

- Mechanical or hydraulic post drivers can be used to pound the posts directly into the ground.
- This technique is efficient and can be suitable for certain soil types.

Rock Socks or Gravel:

- In areas with well-draining soil, installing posts on a bed of rocks or gravel can provide stability and drainage.

Cement Grout Injection:

- For existing holes, injecting cement grout into the soil around the post can improve stability.

Before choosing a specific technique, it's essential to consider local building codes, soil conditions, and the intended use of the pole barn. Consulting with a structural engineer or a professional experienced in post-frame construction can help ensure that the chosen method meets safety and code requirements.

Erecting the Structure

Assembling the Frame

With a solid foundation beneath us, it's time to turn our gaze upwards to the creation of the skeletal framework - the backbone of your barndominium. Much like a skeleton provides structure and shape to the body, the frame of your barndominium will define its form and provide support for the walls, floors, and roof.

Traditional Stick Built Construction:

Assembling the frame is a process that requires precision and attention to detail. Start by laying out the bottom plates of your exterior walls according to your floor plan and secure them to the foundation. These will serve as the blueprint for your wall frames.

Next, construct each wall section flat on the ground, placing studs at 16-inch intervals. The studs, long vertical pieces of lumber, are the bones that give the wall its strength. Crown each wall section with a top plate to tie the studs together and provide a solid attachment point for the roof structure.

Once a wall section is assembled, raise it into position on the bottom plate and secure it in place. Repeat this process for each exterior wall, carefully aligning each section with your floor plan.

Post Frame Pole Barn Construction:

Attaching wall purlins to the posts in a post-frame building is a crucial step in the construction process. Purlins are horizontal members that support the walls and help distribute the load evenly. Here's a general guide on how to attach wall purlins to the posts:

Steps:

Layout:

- Begin by marking the locations where the wall purlins will be attached to each post. The spacing between purlins depends on your building design and local building codes.

Preparation:

- Ensure that the posts are plumb and in the correct positions according to your building design.

Measuring and Cutting Purlins:

- Measure and cut the wall purlins to the appropriate length. Use a level or a straight edge to ensure the cuts are square.
- Positioning Purlins:
- Place the first purlin at the desired height on the posts. You may use temporary braces or clamps to hold it in place while attaching.

Fastening:

- Use the appropriate fasteners (screws, nails, or bolts) to attach the purlins to the posts.
- Ensure that the fasteners are driven straight and penetrate the purlin and the post securely.

Spacing:

- Install additional purlins, maintaining the specified spacing between them. The spacing is usually determined by the design of the building and the type of materials being used.

Leveling:

- Periodically check the level of the purlins to ensure that they are horizontal. Adjust as needed by shimming or repositioning.

Securing Temporary Bracing (if needed):

- If your building design requires temporary bracing during construction, ensure that it is securely in place. Temporary bracing helps maintain the structural integrity of the building until the roof is installed.

Continue Installation:

- Repeat the process for each set of wall purlins, working your way around the building.
- Inspection:
- Inspect the attached purlins for proper alignment, levelness, and secure attachment. Make any necessary adjustments.

Final Checks:

- Before moving on to the next phase of construction, perform a final check to ensure that all wall purlins are securely attached, level, and aligned.

Always follow the building plans and specifications, and adhere to local building codes and regulations. If you are unsure or need guidance, it's advisable to consult with a professional engineer or builder experienced in post-frame construction.

Installing the Walls:

> ***NOTE: For a post-frame building, you will install the roof before moving on to the interior walls.**

As the exterior walls stand firm, the shape of your barndominium begins to take form. With this outline established, we can proceed to install the interior walls. These walls are not just partitions to delineate rooms but also serve as additional support for the structure.

Use your floor plan as a guide to lay out the bottom plates for your interior walls, taking care to ensure accurate placement. Construct and raise each interior wall section as you did with the exterior walls.

As you install each wall, take a moment to appreciate the transformation taking place. What was once an open space is now a labyrinth of rooms and corridors, each with its own purpose and personality.

Securing the Joists

With the walls in place, we can now lay the groundwork for the second story or roof, by installing the joists. Joists are horizontal beams that span the width of the building, providing a base for the floor or roof decking.

Start by installing a rim joist around the exterior of the building, which provides a secure edge for the joist ends. Next, position your joists at 16-inch intervals across the width of the building, securing each end to the rim joist.

As you secure each joist, you're creating a sturdy platform for the next phase of construction - the second-story floor or roof. It's a moment of anticipation, a promise of what's to come.

In essence, erecting the structure of your barndominium is a process of transformation. From the raw materials of lumber and nails, a structure emerges that is both strong and beautiful. It's the realization of your floor plan in three dimensions, a tangible representation of your dream home. With each piece of lumber that's cut and nailed into place, your vision is one step closer to becoming a reality.

Installing the Roof

Choosing Roofing Material

Akin to selecting the perfect hat, the choice of roofing material for your barndominium is a critical decision, one that balances aesthetics, durability, and cost. The roof is your home's first line of defense against the elements, and its material can greatly impact the longevity and maintenance of your barndominium.

Metal roofing is a popular choice for barndominiums, known for its durability and low maintenance requirements. It's available in a variety of colors and styles, allowing you to match it with the overall design of your home. Alternatively, asphalt shingles are a cost-effective option, offering a traditional look and a wide range of color choices.

Installing Trusses

Once you've settled on your roofing material, it's time to commence the installation of the trusses - the triangular framework that supports the roof. While it may be tempting to think of this step as a simple task, in reality, it's a high point of your construction project, both literally and figuratively.

Truss installation requires precision and safety precautions. Each truss should be lifted carefully and secured in its precise location according to your roof design. It's a task that requires teamwork, coordination, and a respect for the laws of physics.

Applying Sheathing

With the trusses standing tall, the next step is to apply the sheathing. Think of sheathing as the canvas upon which the masterpiece of your roof will unfold. It provides a flat, solid surface onto which your chosen roofing material will be installed.

Typically made of large panels of plywood or OSB (oriented strand board), the sheathing is attached to the trusses, covering the entire roof. This step requires accuracy to ensure the panels fit together snugly, creating a continuous, seamless surface.

Attaching Shingles or Metal Panels

As the sheathing lays in wait, the moment has arrived to crown your barndominium with its final layer - the shingles or metal panels. This step is the culmination of your roofing project, the point where your choice of roofing material comes to life.

If you've chosen metal roofing, the panels are laid over the sheathing and secured with screws. The installation starts from the bottom and works up towards the peak of the roof, with each successive panel overlapping the one below it to ensure a watertight seal.

If asphalt shingles are your choice, they are attached with roofing nails, starting from the bottom edge of the roof and working upwards. Each row of shingles overlaps the one below, much like the scales on a fish, effectively shedding water away from the roof.

This stage of roof installation is like the final brushstrokes on a painting, the last notes of a symphony. It's where your vision for your barndominium's exterior becomes a tangible reality, a sight to behold against the open sky. As the last shingle or metal panel is secured, take a moment to stand back and admire the progress you've made. Your barndominium is now one step closer to being a weather-tight, cozy dwelling, ready to welcome you home.

Weatherproofing Your Barndominium

With your barndominium now standing tall, it's time to turn your focus to a crucial aspect of the building process - weatherproofing. This involves creating a weather-resistant barrier between the elements and your home's interior. Think of it as dressing your barndominium in a coat, designed to keep it dry and comfortable, regardless of what Mother Nature throws its way.

Applying Exterior Sealant

The first layer of this protective coat is the application of an exterior sealant. This sealant is a special type of paint or primer that's designed to repel water, resist mold, and provide a durable, long-lasting finish.

Picture yourself as an artist, applying the first layer of paint to a canvas. With each stroke, you're not just adding color, you're also protecting the canvas from damage. When applying the sealant, start at the top of your walls and work your way down, ensuring every nook and cranny is covered.

Installing Siding

With the sealant in place, you're ready to add the next layer of protection - the siding. Siding serves as the outer shell of your barndominium, shielding it from wind, rain, and other weather conditions.

Metal siding, with its durability and low maintenance, is a popular choice for barndominiums. Much like a suit of armor, it provides a sturdy shield for your home, while also adding to its visual appeal. When installing the siding, start at the bottom and work your way up, overlapping each panel to create a waterproof barrier.

Adding Insulation

Next in the line of defense is insulation. While commonly associated with maintaining comfortable indoor temperatures, insulation also plays a key role in weatherproofing your home. By creating a barrier against heat flow, insulation helps keep your home warm in winter and cool in summer, reducing the strain on your HVAC system and saving on energy costs.
Think of insulation as a cozy blanket wrapped around your home. It helps keep the warmth in during the chilly winter months and blocks out the heat during the sweltering summer. When installing insulation, be sure to cover all areas, including the walls, ceilings, and floors, for maximum effectiveness. ** We will delve into insulation much deeper in Chapter 5

Fitting Weather Stripping

The final touch in weatherproofing your barndominium is fitting weather stripping. This involves applying special strips around doors and windows to seal off any gaps that might let in drafts, dust, or moisture.

Imagine you're tucking in the edges of a bedsheet, ensuring a snug fit with no loose corners. That's essentially what you're doing when you install weather stripping. By sealing off these gaps, you're not just improving the comfort of your home, you're also helping to reduce energy consumption and prevent damage from moisture.

With this, your barndominium is now fortified against the elements, dressed in its coat of armor. It's prepared to stand strong and cozy, come rain or shine, providing a safe and comfortable haven for you and your loved ones. As you step back and admire the weatherproofed shell of your barndominium, you can be proud of the progress you've made. But there's still work to be done. Next, we'll be stepping inside to continue our build, creating the inner beauty of your barndominium - the interior construction.

Chapter 5

Getting Your Building Insulated

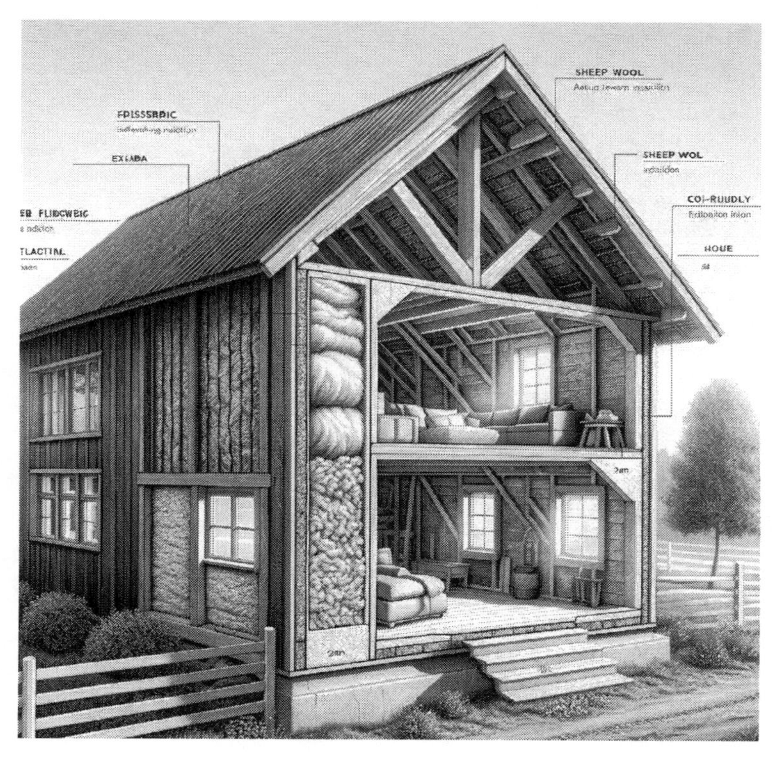

In the quest for energy efficiency and environmental sustainability in the construction industry, the selection of the right insulation material is a critical decision. Four commonly used insulation materials—open-cell spray foam, closed-cell spray foam, fiberglass, and mineral wool—each possess unique characteristics that cater to specific needs and considerations. This chapter delves into a comprehensive comparison of these insulation materials, considering factors such as composition, density, thermal performance, environmental impact, and suitability for various applications.

Composition: Unraveling the Fibers and Cells

Open Cell Spray Foam:

Open-cell spray foam insulation is characterized by a structure where the cells are not fully encapsulated. Composed of polyurethane foam, it expands upon application, creating a semi-rigid structure. The open nature of the cells lends itself to a more breathable material, allowing moisture to pass through. This characteristic makes it particularly suitable for interior applications where moisture management is not a primary concern.

Closed Cell Spray Foam:

In contrast, closed-cell spray foam insulation boasts a more rigid and dense composition. The cells in closed-cell foam are completely encapsulated, resulting in a compact and impermeable structure. This density contributes to its higher insulating properties and makes it an effective air and moisture barrier. Closed-cell foam is often preferred in applications where a robust moisture barrier is essential, such as in exterior insulation.

Fiberglass Insulation:

Fiberglass insulation, a stalwart in the insulation realm, is manufactured from thin fibers of glass that are woven together. The flexibility of these glass fibers allows for easy installation in various spaces. The material is lightweight, providing a different dynamic compared to the foam counterparts. However, fiberglass does not create a complete air or moisture barrier due to its inherent structure.

Mineral Wool Insulation:

Derived from minerals like basalt or recycled slag, mineral wool insulation strikes a balance between flexibility and density. Its composition results in a material that is resistant to fire and can withstand high temperatures. Mineral wool insulation is inherently non-combustible, making it a preferred choice in applications where fire resistance is a primary concern.

Density and R-Value: Navigating the Thermal Landscape

Open Cell Spray Foam:

The density of open-cell spray foam is lower compared to closed-cell foam. While it offers respectable thermal performance, its lower density translates to a lower R-value per inch. The lower R-value can be a factor in determining the thickness of the insulation needed to achieve the desired thermal resistance.

Closed Cell Spray Foam:

Closed-cell spray foam, with its higher density, presents a more substantial R-value per inch. This characteristic makes it particularly advantageous in scenarios where space is limited, and a higher insulation value is required. The rigid nature of closed-cell foam also adds structural strength to the building.

Fiberglass Insulation:

Fiberglass insulation falls within a moderate density range, providing a respectable R-value. The R-value of fiberglass is influenced by factors such as the thickness of the insulation and installation techniques. While fiberglass insulation can effectively insulate spaces, it may require thicker applications to achieve comparable R-values to some foam insulation materials.

Mineral Wool Insulation:

With a density similar to fiberglass, mineral wool insulation offers a competitive R-value. Its ability to withstand high temperatures without compromising performance adds to its appeal, especially in applications where exposure to heat is a concern.

Air and Moisture Barrier: Guarding Against Infiltration

Open Cell Spray Foam:

Open-cell spray foam acts as an air barrier, preventing the infiltration of air through the building envelope. However, it is not inherently a moisture barrier. While it can resist moisture to some extent, it may not provide the same level of protection as closed-cell foam in humid or wet conditions.

Closed Cell Spray Foam:

The encapsulated cells in closed-cell spray foam create a formidable barrier against both air and moisture. This characteristic makes it a preferred choice in areas prone to high humidity, extreme weather conditions, or applications where a complete moisture seal is essential.

Fiberglass Insulation:

Fiberglass insulation allows air to pass through, and its effectiveness as an air barrier depends on proper installation techniques and additional sealing materials. While it may resist moisture to some extent, it does not function as a complete moisture barrier.

Mineral Wool Insulation:

Mineral wool insulation offers some resistance to air movement but is not as effective as closed-cell spray foam. Its ability to handle moisture is influenced by factors such as the density and specific formulation of the material.

Installation and Application: Tailoring to Project Needs

Open Cell Spray Foam:

Open-cell spray foam is relatively cost-effective compared to closed-cell foam, making it an attractive option for projects with budget constraints. Its flexibility and breathability make it suitable for interior applications, such as insulating walls and ceilings. However, it may not be the ideal choice for exterior applications or areas with high moisture levels.

Closed Cell Spray Foam:

While closed-cell spray foam comes at a higher cost, its denser composition and moisture resistance make it suitable for a broader range of applications. It is commonly used in exterior insulation, roofing projects, and areas with high humidity or moisture exposure.

Fiberglass Insulation:

Fiberglass insulation is known for its ease of installation, making it a popular choice for attics, walls, and other spaces. Its flexibility allows for easy adaptation to different structural configurations. The cost-effectiveness of fiberglass insulation further contributes to its widespread use.

Mineral Wool Insulation:

The installation of mineral wool insulation is comparable to that of fiberglass. It can be used in similar applications, including walls, attics, and floors. However, specific handling precautions may be necessary due to its density and potential for skin irritation.

Environmental Impact: Balancing Efficiency and Eco-Friendliness

Open Cell and Closed Cell Spray Foam:

Both open-cell and closed-cell spray foams may contain chemicals that contribute to greenhouse gas emissions during production. Closed cell foam, with its higher density and often requiring more resource-intensive manufacturing processes, may have a higher environmental impact. Evaluating the specific formulations and considering the long-term energy savings can help in assessing the overall environmental impact.

Fiberglass Insulation:

Fiberglass insulation is manufactured from recycled glass and sand, contributing to its relatively lower environmental impact. Additionally, fiberglass insulation can be recycled, reducing the demand for new raw materials.

Mineral Wool Insulation:

Mineral wool insulation is made from abundant and recyclable materials, presenting a lower environmental impact compared to some other insulation types. The use of recycled materials in its production further enhances its eco-friendly profile.

Fire Resistance: Safeguarding Against Flames

Open Cell Spray Foam:

While open-cell spray foam offers some fire resistance, it is not as inherently fire-resistant as closed-cell foam. Its use may be limited in applications where stringent fire codes must be met.

Closed Cell Spray Foam:

Closed-cell spray foam excels in terms of fire resistance due to its compact and rigid structure. It is often chosen for applications where meeting stringent fire codes is imperative.

Fiberglass Insulation:

Fiberglass insulation exhibits good fire resistance properties, adding a layer of safety to the building. It is inherently non-combustible, contributing to its widespread use.

Mineral Wool Insulation:

Mineral wool insulation is renowned for its excellent fire resistance. Its ability to withstand high temperatures without compromising its structural integrity makes it a preferred choice in applications where fire safety is a top priority.

Conclusion: Tailoring the Insulation Solution

In the intricate tapestry of building construction, the choice of insulation material is a pivotal thread that weaves together considerations of cost, efficiency, sustainability, and safety. Open-cell spray foam, closed-cell spray foam, fiberglass, and mineral wool each offer a unique set of characteristics, catering to specific project requirements and constraints.

The decision-making process should involve a careful evaluation of factors such as the desired R-value, budget limitations, environmental considerations, and the specific challenges posed by the project's location and conditions. Open cell spray foam, with its cost-effectiveness and flexibility, may find a niche in interior applications where moisture management is not a primary concern. Closed-cell spray foam, with its higher density and superior moisture resistance, may be the preferred choice for exterior applications or projects in humid climates.

Fiberglass insulation, a stalwart in the insulation landscape, provides a balance of cost-effectiveness and ease of installation. Its recyclability further enhances its appeal from an environmental perspective. Mineral wool insulation, with its fire resistance and recyclable composition, emerges as a robust option in applications where safety and sustainability are paramount.

In the dynamic realm of building science, the comparative analysis of insulation materials is a guidepost rather than a rigid prescription. As technology advances and new formulations emerge, the landscape of insulation options continues to evolve. The prudent approach involves a holistic consideration of the project's unique requirements, striking a balance between efficiency, cost-effectiveness, and environmental responsibility. Ultimately, the insulation choice becomes an integral part of the larger narrative of constructing spaces that are not only energy-efficient but also resilient and sustainable for generations to com

We would love your Feedback to help us make someones Barndo dreams come true

"Kindness is the language which the deaf can hear and the blind can see."

- Mark Twain

Hello there, fellow builders and dreamers,

We're thrilled that you've joined us in the exciting journey of exploring "Building Your Barndominium: A Comprehensive Guide to DIY Barn Home Building, Creating Rural Living Home Designs, and Unique Barndo Spaces on Any Budget" by BarnDream Builders.

Our Mission:

Our mission is simple - to make the world of barndominium building accessible to everyone. Every action we take is guided by this mission. To achieve it, we need your support.

Now, we have a special request that will make a real difference in someone else's life. Are you ready to extend a helping hand to a fellow builder you've never met?

How You Can Help:

People often judge a book by its reviews, and your honest opinion can make a world of difference to someone considering this book. Your review could:

Inspire one more small business to support its community.

Empower one more entrepreneur to provide for their family.

Assist one more employee in finding meaningful work.

Enable one more client to transform their life.

Fulfill one more dream.

The Power of Your Review:

Leaving a review costs you nothing but a few moments of your time, yet it can potentially change the trajectory of someone else's life. It's a simple act of kindness that can have a profound impact.

How to Leave a Review:

To help this aspiring builder without even meeting them, please leave a review by following this link:

Join the Club:

If the idea of helping a fellow builder, even if they remain a stranger, resonates with you, then you're the kind of person we admire. Welcome to our club of compassionate builders.

A Heartfelt Thank You:

Your willingness to help is deeply appreciated. You're not just contributing to the success of this book; you're making a positive impact on someone's journey toward their dreams.

With gratitude,

BarnDream Builders

Chapter 6

Utility Requirements

Your Barndominium is going to be unique to you and your family. Depending on the location of your build, several utility questions need to be addressed. Will you have access to natural gas? Will you be needing a propane tank on your property? Are you going to go with an all-electric set-up? Do you plan on using solar panels or renewable energy? Will you need to engineer a septic field?

Let us touch on these to try and provide some direction with your plans.

Understanding Electrical Service Requirements:

As the construction industry evolves, the requirements for electrical installations in new residential constructions have become more stringent and sophisticated. Modern homes are now designed to accommodate a plethora of electronic devices, energy-efficient systems, and smart home technologies. This necessitates a comprehensive understanding of electrical service requirements to ensure safety, efficiency, and compliance with building codes.

1. Basic Electrical Service Requirements:

When constructing a new residential property that relies solely on electricity for all its systems—such as heating, cooling, cooking, and water heating—it's crucial to determine the appropriate size for the electrical service.

- **Single-Family Homes:** For a typical single-family home that utilizes only electricity, a 200-amp electrical service is commonly recommended. This allows ample capacity to meet the demands of modern appliances and electronics without overloading the system. However, larger homes or those with extensive electrical requirements may necessitate a 400-amp service or even higher.

2. Electrical Service with Natural Gas or Propane:

When a residential property incorporates both electrical and gas systems, the electrical service requirements can be adjusted based on the energy demands of each system.

- **Dual-Fuel Heating Systems:** Homes equipped with dual-fuel heating systems, which utilize both electricity and natural gas or propane, will generally have a reduced electrical demand compared to all-electric homes. In such cases, a 150-amp to 200-amp electrical service might suffice for most single-family homes, as the heating load is shared between the gas and electric systems.

- **Gas Cooking and Water Heating:** If the house uses gas for cooking and water heating while relying on electricity for other systems, a 150-amp electrical service is typically adequate. Gas-powered appliances reduce the overall electrical demand, allowing for a smaller electrical service size.

3. Considerations for Future Expansion:

When determining the electrical service size for new residential construction, it's essential to consider future expansion and potential increases in electrical demand. Incorporating additional circuits or planning for the integration of electric vehicle (EV) charging stations, solar panels, or home automation systems can influence the recommended service size.

Selecting the appropriate electrical service size for new residential construction involves careful planning and consideration of the home's specific energy requirements. Whether the home relies solely on electricity or incorporates both electrical and gas systems, adhering to local building codes and consulting with electrical professionals is essential to ensure a safe and efficient electrical installation. By understanding these requirements and anticipating future needs, homeowners and builders can create homes that are equipped to handle the demands of modern living while promoting energy efficiency and sustainability.

Research your local electrical service provider. Explore your options when it comes to where to put the service panel on your building. Will you need to trench to your building? How far will your service provider allow your drop to be from the transformer/pole? Once your service is run, where will your service panel be inside of your building? Always try to keep in mind that most Barndominiums do not have basements. A suitable utility room will need to be included in your floor plan.

Natural gas / Propane:

When planning a new construction where natural gas or propane will be utilized as a utility, it is imperative to adhere to specific construction requirements to ensure safety, efficiency, and compliance with local building codes. Herein, we outline the primary requirements for such constructions:

1. Safety Precautions:

- **Ventilation:** Proper ventilation is paramount when dealing with natural gas or propane. Adequate air circulation ensures the safe dispersion of any potential leaks, preventing the accumulation of dangerous concentrations.
- **Gas Detection Systems:** Install gas detectors in areas where gas appliances are present, such as kitchens or

utility rooms. These detectors can alert occupants to the presence of gas leaks, providing an early warning to evacuate if necessary.

- **Emergency Shut-off Valves:** Integrate emergency shut-off valves near gas meters or at strategic locations within the home. These valves allow for the immediate cessation of gas flow in case of emergencies.

2. Infrastructure Requirements:

- **Gas Lines:** All gas lines must be constructed with durable, corrosion-resistant materials such as steel or flexible copper. Proper sealing and joint connections are crucial to prevent leaks.
- **Pressure Regulation:** Depending on the local supply, it may be necessary to incorporate pressure regulators to ensure that the gas is delivered at a safe and consistent pressure throughout the home.
- **Pipe Routing:** Gas pipes should be routed in a manner that minimizes potential damage from external factors and allows for easy access for maintenance and inspection.

3. Appliance Installation and Venting:

- **Certified Appliances:** Only install gas appliances that are certified and approved for use in residential settings. Ensure that these appliances are installed according to manufacturer guidelines and local codes.
- **Proper Venting:** Gas appliances produce combustion by-products that must be safely vented to the exterior of the home. Install appropriate venting systems, such as flues or chimneys, for each appliance to prevent the accumulation of harmful gasses like carbon monoxide.

4. Compliance and Inspection:

- **Permitting:** Obtain all necessary permits before commencing construction. Local authorities may require inspections at various stages to ensure compliance with building and safety codes.
- **Certified Professionals:** Engage licensed and certified professionals for the installation of gas utilities and appliances. Regular inspections and maintenance by qualified technicians are essential to ensure continued safety and efficiency.
- **Code Adherence:** Familiarize yourself with local, state, and national codes pertaining to gas installations. Adherence to these codes not only ensures safety but also avoids potential legal complications.

5. Educational Resources:

- **Occupant Training:** Provide occupants with information and training on the safe use and maintenance of gas appliances. This includes understanding how to operate appliances, recognizing signs of potential issues, and knowing what to do in case of emergencies.
- **Documentation:** Maintain comprehensive documentation of all installations, inspections, and maintenance activities related to gas utilities. This documentation can be invaluable for future reference and may be required for insurance purposes or property transactions.

Incorporating natural gas or propane utilities into a new construction requires meticulous planning, adherence to safety protocols, and compliance with relevant regulations. By prioritizing safety, utilizing qualified professionals, and maintaining diligent oversight throughout the construction process, homeowners can enjoy the benefits of these energy sources with confidence and peace of mind.

Solar Panels / Renewable Energy:

As sustainability becomes increasingly important, many homeowners are considering integrating renewable energy solutions, such as solar panels, into their new construction homes. This subchapter delves into the decision-making process, comparing upfront costs with potential long-term cost savings associated with renewable energy adoption.

1. Initial Investment: Upfront Costs

- **Solar Panels:** The upfront costs for installing solar panels can be substantial, encompassing equipment, installation labor, permits, and potentially, energy storage solutions like batteries.

- **Renewable Energy Systems:** Other renewable energy systems, such as wind turbines or geothermal heat pumps, also involve significant upfront investments in equipment and installation.

2. Financial Incentives and Tax Credits

- **Solar Incentives:** Various financial incentives, including federal tax credits, state rebates, and utility incentives, can offset a portion of the initial investment in solar panels, reducing the effective upfront cost.
- **Other Renewable Incentives:** Similar incentives may be available for other renewable energy systems, enhancing their cost-effectiveness.

3. Long-Term Cost Savings: Operational Savings

- **Energy Savings:** One of the primary benefits of renewable energy systems is the potential for significant energy savings over their operational lifespan. Solar panels, for example, can substantially reduce or even eliminate monthly electricity bills, leading to substantial savings over time.
- **Energy Independence:** By generating their own renewable energy, homeowners can reduce their reliance on the grid, potentially insulating themselves from rising utility costs.

4. Return on Investment (ROI)

- **Payback Period:** Calculating the payback period— the time it takes for the cumulative energy savings to offset the initial investment— provides insights into the financial viability of renewable energy systems. Generally, the shorter the payback period, the more attractive the investment.

- **ROI Considerations:** Factors such as the cost of electricity, system efficiency, maintenance expenses, and potential future energy price increases can influence the ROI of renewable energy investments.

5. Lifecycle Costs and Maintenance

- **Maintenance Costs:** While renewable energy systems often have lower operational costs than traditional energy sources, they may require periodic maintenance or component replacements, which can impact long-term costs.
- **System Lifespan:** Understanding the expected lifespan of renewable energy systems and factoring in potential future upgrades or replacements is crucial for accurate long-term cost projections.

6. Environmental and Sustainability Considerations

- **Environmental Benefits:** Beyond financial considerations, renewable energy adoption contributes to environmental sustainability by reducing greenhouse gas emissions, minimizing reliance on fossil fuels, and promoting clean energy production.
- **Resale Value:** Homes equipped with renewable energy systems may command higher resale values, reflecting their energy efficiency, sustainability features, and potential cost savings for future owners.

Choosing to integrate solar panels or other renewable energy systems into a new construction home involves balancing upfront costs with potential long-term cost savings, environmental benefits, and lifestyle considerations. While the initial investment may be substantial, financial incentives, energy savings, and other factors can enhance the overall cost-effectiveness of renewable energy adoption. Careful evaluation of upfront costs, ongoing operational expenses, and potential financial incentives will empower homeowners to make informed decisions that align with their sustainability goals and financial objectives.

Septic and Waste

The majority of Barndominium homes will be built in rural areas. It is important to decide how you will deal with the drain and waste from your home. Installing a septic system is a significant undertaking for any new construction home. Proper planning ensures that the system functions efficiently, meets local regulations, and lasts for many years without issues. Here's a guide to the essential planning steps:

1. Site Assessment and Soil Testing

- **Site Evaluation:** Before anything else, a thorough site assessment is crucial. Consider factors like topography, proximity to water sources, and the location of the home on the property.
- **Soil Testing:** Different soil types have varying rates of water absorption. A perc test (percolation test) is performed to determine how quickly the soil can absorb water. This test will guide the design and sizing of the septic field.

2. Regulatory Compliance and Permits

- **Local Regulations:** Check with local health departments or authorities to understand regulations, setback requirements, and any permits needed for septic installations.
- **Permitting:** Secure all necessary permits before beginning any work. This often involves submitting site plans, soil test results, and septic system designs for approval.

3. System Design

- **Tank Size and Type:** Based on the anticipated wastewater volume and soil test results, determine the appropriate size and type of septic tank (e.g., concrete, fiberglass).
- **Distribution and Absorption:** Design the septic field layout considering the soil's percolation rate. This may involve trenches, beds, or other configurations to ensure even distribution of effluent and effective filtration.

4. Construction Logistics

- **Access and Excavation:** Ensure there's adequate access for machinery and trucks required for excavation and installation. Consider any existing structures, trees, or utilities that may impact the installation.
- **Safety Measures:** Establish safety protocols, especially for excavation work. This includes ensuring proper shoring, marking utilities, and using appropriate personal protective equipment.

5. Budget and Financing

- **Cost Estimation:** Get detailed quotes from contractors for the entire installation process, including excavation, materials, labor, and any required permits or inspections.
- **Financing Options:** Explore financing options if needed, such as construction loans or home improvement loans specifically for septic system installations.

6. Future Maintenance Considerations

- **Access Points:** Ensure the septic tank and field have adequate access points for inspection, maintenance, and pumping.

- **Maintenance Schedule:** Develop a maintenance schedule based on the system's size, usage, and local recommendations. Regular inspections and periodic pumping are essential for long-term functionality.

7. Environmental and Health Considerations

- **Environmental Impact:** Consider the environmental impact of the septic system, especially if the property is near sensitive areas like wetlands or water bodies.
- **Health and Safety:** Ensure that the septic system is designed and installed to prevent health risks, such as groundwater contamination or surface water runoff.

By meticulously planning each phase of the septic system installation, homeowners can ensure a reliable, efficient, and environmentally responsible wastewater management solution for their new construction home. Collaboration with experienced professionals, adherence to local regulations, and proactive maintenance will further enhance the system's longevity and performance.

Fresh Water / Well

Installing a freshwater well for a new construction home involves careful planning and adherence to regulatory guidelines to ensure a safe and sustainable water supply. Here's a comprehensive guide to the essential planning steps:

1. Site Assessment and Hydrogeological Study

- **Site Evaluation:** Conduct a thorough site assessment to identify potential well locations. Consider factors such as topography, proximity to contamination sources, and accessibility for drilling equipment.

- **Hydrogeological Study:** Perform a hydrogeological study to understand the groundwater conditions, aquifer characteristics, and potential yield of the well. This study will guide the well's design and placement.

2. Regulatory Compliance and Permits

- **Local Regulations:** Familiarize yourself with local, state, and federal regulations governing well drilling, water quality, and setback requirements from potential contamination sources.
- **Permitting:** Obtain all necessary permits before commencing drilling activities. This typically involves submitting well construction plans, hydrogeological study findings, and other relevant documentation for regulatory approval.

3. Well Design and Construction

- **Well Depth and Diameter:** Based on the hydrogeological study, determine the optimal well depth and diameter to access a sustainable water supply. The well design should consider the anticipated water demand and potential future needs.
- **Materials and Construction Methods:** Select appropriate materials (e.g., casing, screen) and construction methods to ensure the well's integrity, prevent contamination, and maximize water yield. Follow industry best practices and regulatory guidelines for well construction.

4. Water Quality Testing

- **Initial Testing:** Conduct baseline water quality testing before using the well to establish a reference for future comparisons. This testing will identify any existing contaminants and inform necessary water treatment measures.

- **Routine Monitoring:** Implement a routine water quality monitoring program to ensure ongoing compliance with drinking water standards and early detection of potential issues.

5. Infrastructure and Accessibility

- **Wellhead Protection:** Establish a wellhead protection area to prevent potential contamination from surface activities, such as landscaping, storage, or chemical use.
- **Accessibility:** Ensure the well is easily accessible for maintenance, testing, and potential future upgrades. Consider installing protective measures, such as well caps or fences, to prevent unauthorized access and protect the wellhead.

6. Budget and Financing

- **Cost Estimation:** Obtain detailed cost estimates from well drilling contractors, including drilling, materials, permits, and initial water quality testing.
- **Financing Options:** Explore financing options, such as construction loans or home improvement loans, to cover the upfront costs of well installation.

7. Maintenance and Monitoring

- **Well Maintenance:** Develop a regular maintenance schedule to inspect the well, monitor water levels, and address any issues promptly. Well components may require periodic replacement or repair to ensure optimal performance.

- **Water Use Monitoring:** Implement water use monitoring to track consumption patterns, identify potential leaks or inefficiencies, and optimize water conservation practices.

By carefully planning each phase of the freshwater well installation, homeowners can establish a reliable and sustainable water supply for their new construction home. Collaboration with experienced professionals, compliance with regulatory requirements, and proactive maintenance will ensure the well's longevity and the delivery of safe, high-quality drinking water for years to come.

The successful planning and construction of a new home necessitates a comprehensive understanding and integration of various utility systems. From the crucial elements of electrical wiring, plumbing, and heating, ventilation, and air conditioning (HVAC) systems to considerations of water supply, waste management, and internet connectivity, each utility system plays a vital role in shaping the functionality, comfort, and sustainability of a home. As homeowners embark on the exciting journey of building their dream residence, a thoughtful and well-coordinated approach to these utility systems becomes imperative. By recognizing the interdependence of these components and staying abreast of innovative technologies and eco-friendly solutions, individuals can ensure that their new homes not only meet their immediate needs but also contribute to a resilient and efficient living space for years to come.

Chapter 7

Interior Construction: Framing Your Vision

Have you ever watched an artist sketching a portrait? They start with a few simple lines, marking out the boundaries of the face. Gradually, they add more lines, defining the contours, shapes, and features. Bit by bit, a human face emerges from the blank canvas, each line a testament to the artist's vision and skill. This is the magic of framing - the art of bringing an image to life, one line at a time.

Similar to an artist's sketch, the interior framing of your barndominium serves as the blueprint upon which the details of your home will take shape. It's the skeletal structure that partitions your open barndominium shell into rooms and corridors, each serving a unique function. As we step into this chapter, we'll delve into the process of interior framing, transforming your barndominium shell from an open canvas to a labyrinth of spaces, each with its unique purpose and personality.

Framing the Interior

Planning Room Layouts

Just as an artist uses a pencil to outline their drawing before they begin, your first step in interior framing is planning your room layouts. This is when your floor plan transitions from a two-dimensional drawing to a three-dimensional space.

Take a step back and visualize your barndominium as a completed home. Imagine walking through the front door, moving from room to room. Think about the function of each space, the flow between them, and how the sunlight changes throughout the day. Consider the placement of doors and windows and how they will affect movement and light.

Installing Stud Walls

With your room layout planned, it's time to start building your stud walls. This process is similar to drawing a grid on a blank canvas before you start painting. Each line of the grid is a guide for the placement of an element in the painting. In the same way, each stud wall you install is a guide for the placement of a room in your barndominium.

Stud walls are built flat on the ground and then tilted up and secured in place. This method allows for a smoother and safer installation process. Start by cutting your studs and plates to the correct lengths, then lay out the pieces on the ground, according to your wall plan. Once everything is in place, nail the studs to the plates to create a wall section.

Setting Door Frames

Setting door frames is like inserting a keyhole into a finely crafted wooden box. It requires precision and care, and the result is both functional and aesthetically pleasing.

Start by determining the location of each door according to your floor plan. Cut out the section of the stud wall where the door will go, making sure to leave an extra half-inch on each side for the door jamb. Install a horizontal beam, or header, across the top of the door opening to provide support. Then, install a temporary support stud, or cripple, under the header until the door frame is installed.

Mounting Ceiling Joists

The final step in framing your interior is mounting the ceiling joists. These horizontal beams provide a base for your ceiling surface and also help tie the walls together.

Think of this step as installing the top of a box, giving it rigidity and defining its shape. The joists should be evenly spaced across the width of your barndominium and securely nailed to the top plates of the walls. If you plan to install a second floor, the joists will also provide support for the floor structure.

Like an artist sketching a portrait, framing your barndominium's interior requires precision, patience, and an eye for detail. But the result is worth every effort. With each stud wall and ceiling joist you install, you're bringing your vision to life, transforming an empty shell into a place you can call home. As you step back and admire your handiwork, take a moment to appreciate the beauty of this transformation, and look forward to the next stage of your barndominium-building journey.

Installing Insulation

Choosing Insulation Type

In Chapter 5 we discussed the insulation types and features that each type will provide for your build. As this is a vital element to the longevity and efficiency of your home, we will touch once again on them before getting into actual installation.

When it comes to selecting the right insulation for your barndominium, think of it as wrapping a present destined for a beloved. You'll want to consider its longevity, reliability, and effectiveness. There are several types of insulation to choose from, each with its own set of advantages and appropriate applications.

Fiberglass insulation, available in batts or loose fill, is a common choice due to its ease of installation and cost-effectiveness. It provides decent thermal resistance and is fire-resistant, making it a safe choice for most home applications.

Spray foam insulation, although more costly upfront, offers superior thermal resistance and has the added benefit of sealing small cracks and gaps, improving your barndominium's overall air tightness. This type of insulation can be a smart choice for hard-to-reach areas or for insulating around windows and doors.

Mineral wool insulation, made from spun rock or steel slag, is highly fire-resistant and also offers excellent soundproofing. This can be an excellent choice if you're looking to create a quiet, serene interior environment.

Insulating Walls

The walls of your barndominium are like the layers of a well-crafted sandwich. The bread slices represent the interior and exterior wall surfaces, while the filling in between is your insulation.

Begin the process by measuring the distance between your wall studs, this will determine the width of insulation you'll need. Whether you're using fiberglass batts or spray foam, the goal is to fill the entire cavity between your wall studs, leaving no gaps. This ensures an even thermal barrier, preventing heat transfer through your walls.

When installing fiberglass batts, start at the top of the wall and work your way down, securing the batts in place with staples or insulation hangers. For spray foam, simply spray the foam into the wall cavity, ensuring a complete and even fill.

Insulating Ceilings

Turning your attention upwards, consider the ceilings of your barndominium. Here, insulation serves to keep warm air from escaping through the roof in the winter and prevents hot air from seeping in during the summer.

If you have an accessible attic space, loose-fill fiberglass or cellulose insulation can be an effective and economical choice. These materials can be blown into the attic space to a specified depth, providing a thick layer of insulation.

For finished ceilings or sloped ceilings, consider using spray foam or cut-to-fit batts. As with wall insulation, ensure the entire cavity between the ceiling joists is filled, leaving no gaps.

Insulating Floors

Finally, don't overlook the floors of your barndominium. Floor insulation can make a significant difference in the comfort of your home, especially if you have unheated spaces below or if your home is built over a crawlspace or unheated basement.

For floors, fiberglass batts are a common choice. Install the batts between the floor joists, ensuring a snug fit. If you're using faced batts, the vapor barrier should face upwards, towards the heated space. Secure the batts with insulation hangers or wire supports.

In the case of a concrete slab floor, rigid foam insulation can be installed directly beneath the slab. This not only insulates the floor but can also help reduce potential moisture problems.

In essence, the process of installing insulation in your barndominium is like wrapping your home in a protective envelope. With each piece of insulation placed, you're enhancing the comfort, efficiency, and durability of your home, preparing it to provide a warm and cozy shelter for you and your loved ones. As the final piece of insulation is secured and you wipe the sweat from your brow, pause to appreciate the significance of this moment. You're not just building a house. You're building a sanctuary, a haven, a place where memories will be made and dreams will unfold.

HVAC

A crucial design decision is how you are going to heat and cool your home. This decision will play a huge part in how you lay out and design your rooms and spaces. Electrical and gas requirements must be considered. The need for ductwork or to plan for wall mounted units. The need for a boiler, or a chimney. Let's take a look at some of the options to be considered.

In-Floor Radiant Heat:

Description: In-floor radiant heating involves installing a system of pipes or electric heating elements beneath the floor. Heat radiates from the floor surface, providing even warmth to the space.

Pros:

- Energy-efficient, as heat rises from the floor.
- Even heating distribution.
- No visible radiators or ductwork.

Cons:

- Initial installation costs can be high.
- Slower response time compared to forced-air systems.

Forced Air Heating and Cooling:

Description: Forced air systems use a furnace to heat air, which is then distributed throughout the home via a system of ducts. The same system can be used for air conditioning by adding a cooling component such as an air conditioner or heat pump.

Pros:

- Rapid heating and cooling response.
- Can include air filtration systems.
- Compatible with programmable thermostats.

Cons:

- Ductwork may require maintenance.
- Air distribution may not be as even as radiant heat.

Geothermal Heating and Cooling:

Description: Geothermal systems use the constant temperature of the earth to transfer heat to or from a home. This is achieved through a ground-source heat pump connected to a series of buried pipes or loops.

Pros:

- Highly energy-efficient.
- Lower operating costs over time.
- Environmentally friendly.

Cons:

- High upfront installation costs.
- Requires available land for loop installation.

Ductless Mini-Split Systems:

Description: Ductless mini-split systems consist of an outdoor compressor/condenser unit and one or more indoor air-handling units. These units are connected by refrigerant lines and require no ductwork.

Pros:

- Zoned heating and cooling for better energy efficiency.
- Easy installation without ductwork.
- Can be used for both heating and cooling.

Cons:

- The initial cost can be higher than traditional systems.
- Multiple indoor units may be visible on walls.

Each HVAC system has its own set of advantages and considerations. The choice depends on factors like climate, budget, energy efficiency goals, and personal preferences. It's recommended to consult with HVAC professionals or energy consultants to determine the most suitable system for a specific home.

Electrical and Plumbing Basics

Mapping Electrical Outlets and Switches

Consider the flow of electricity in your barndominium as a network of roads, delivering power to every corner of your home. The first step in setting up this network is mapping out the placement of your electrical outlets and switches. From the kitchen to the bathroom, from the living room to the bedroom, each room has its unique electrical needs.

Start by identifying the key activity areas in each room. For example, in the kitchen, you'll need outlets for appliances like the refrigerator, stove, and possibly a microwave or dishwasher. In the bedroom, think about the placement of the bed and side tables, where you'll likely need outlets for lamps, phone chargers, or an alarm clock.

Consider the height of your outlets as well. Standard outlets are typically 12 inches above the floor, but you might want to install some at counter height in the kitchen or bathroom for convenience. Switches, on the other hand, are usually installed 48 inches above the floor, making them easily accessible.

Running Electrical Wires

Once the placement of outlets and switches is determined, it's time to run the electrical wires. This is akin to laying the roads in your electrical network, connecting all the outlets and switches to your main electrical panel.

Start by drilling holes in the center of the wall studs and floor joists, through which you'll run the wires. Each circuit should have a hot wire, which delivers power to the outlet or switch, a neutral wire, which returns the electrical current to the panel, and a ground wire, which provides a safe path for electricity in case of a fault.

Remember, safety is paramount when dealing with electricity. Always turn off the power at the main panel before working with electrical wires, and consider hiring a licensed electrician to ensure the work is done correctly and safely.

If your Barndo is going to concrete floors, make sure to pre-plan for anything electrical needs under your concrete slab. Flush to floor outlets, or an outlet in a kitchen island will need to be planned before pouring a slab.

Installing Plumbing Lines

Water, much like electricity, is a vital resource that needs to be efficiently distributed throughout your barndominium. The process of installing plumbing lines is similar to setting up your electrical network, but instead of wires, you'll be working with pipes.

First, identify where you'll need plumbing fixtures, such as sinks, toilets, showers, and washing machines. Each of these fixtures will need both a supply pipe, to bring in fresh water, and a drain pipe, to carry away wastewater.

Next, run your supply and drain pipes, connecting each fixture to your main water line and sewer system. Supply pipes are usually smaller and under pressure, while drain pipes are larger and rely on gravity to move the water.

Again, safety and accuracy are crucial. Plumbing must be done correctly to prevent leaks, water damage, and health hazards. Hiring a professional plumber can be a smart move, particularly for complex installations.

As with the electrical lines, with a concrete slab floor, all plumbing runs need to be planned ahead of pouring your slab. In-floor radiant heat is a popular option in Bardo's. The piping for these systems must be installed prior to the floor being poured.

Open floor plans and large open spans might require some plumbing lines to be run under the floor.

Fitting Fixtures

The final step in setting up your electrical and plumbing systems is fitting the fixtures. This involves installing light fixtures, electrical outlets, switches, faucets, showerheads, toilets, and other appliances.

Each fixture needs to be carefully selected to suit the style and functionality of your barndominium. Consider the aesthetics, energy efficiency, and durability of each fixture. For example, LED light fixtures are energy-efficient and long-lasting, while a dual-flush toilet can save water and reduce your utility bills.

When installing these fixtures, precision is key. Electrical fixtures should be securely mounted and properly wired, while plumbing fixtures need to be tightly fitted to prevent leaks.

With these steps, your barndominium is now equipped with the basic utilities, ready to light up and quench its thirst. As you stand in your almost-complete barndominium, you can almost hear the soft hum of electricity, the gentle flow of water, waiting to breathe life into your home. Just a few more steps and your barndominium will be ready to welcome you to a life of comfort and rustic charm.

Choosing and Installing Flooring and Finishes

Selecting Flooring Material

Imagine standing at the edge of a beautiful forest, gazing at the array of trees before you. Each species of tree has its unique characteristics, strengths, and beauty. Similarly, when selecting flooring material for your barndominium, you have a variety of options, each with its distinct advantages and aesthetics.

Hardwood flooring, with its natural beauty and durability, is a popular choice for living areas and bedrooms. It adds warmth and character to any room, and its longevity makes it a cost-effective choice in the long run. Alternatively, ceramic or porcelain tiles are excellent for wet areas like the kitchen or bathroom, owing to their water and stain resistance. For a budget-friendly option, consider laminate or vinyl flooring, which offers a wide range of designs and is easy to install and maintain.

Preparing Subfloor

The process of preparing the subfloor is akin to a chef meticulously prepping their ingredients before starting to cook. It involves creating a clean, level, and stable surface for your flooring material.

Begin by removing any debris, old adhesive, or remnants of previous flooring. Next, check the levelness of the floor using a long level or a straight board. If you find any low spots, fill them with a self-leveling compound. For high spots, sand them down until they are level with the rest of the floor.

Laying Flooring

With your subfloor prepped and ready, the stage is set to start laying your chosen flooring, much like a gardener carefully planting seeds in a well-prepared garden bed.

If you're installing hardwood, start at the longest, most visible straight wall and work your way across the room, ensuring the boards are straight and secure. For tiles, start in the center of the room and work your way outwards, using spacers to ensure even grout lines. If you're using laminate or vinyl planks, start at one wall and work towards the opposite wall, carefully cutting the planks to fit as you go.

Applying Wall Finishes

Turning our attention to the walls, applying the finishes is like a painter adding the final layers of color to a masterpiece.

Drywall is a common wall finish due to its smooth surface and ease of installation. Once the drywall is in place, you can further enhance it with paint, wallpaper, or even a textured plaster finish. To add character and warmth to your barndominium, consider incorporating a feature wall using reclaimed wood or stone veneer.

Installing Cabinetry and Countertops

The installation of cabinetry and countertops is like placing the furniture in a dollhouse. It's a process that requires precision and care, as these elements are both functional and integral to the aesthetics of your kitchen and bathrooms.

When installing cabinets, start with the upper ones, ensuring they are level and securely fastened to the wall studs. Next, install the lower cabinets, taking care to align them with the uppers. Finally, carefully measure, cut, and attach the countertops, making sure they are level and securely attached to the cabinets.

Painting and Decorating

The final phase in your interior construction process is painting and decorating, akin to adding icing and decorations to a cake.

When painting, start with the ceiling, followed by the walls, and finally the trim. Choose colors that reflect your style and complement the overall design of your barndominium.

Decorating involves adding the finishing touches that make your house feel like a home. This could include hanging artwork, placing rugs, arranging furniture, and adding personal mementos.

It's your chance to infuse your personality into your space, creating an environment that is uniquely yours.

With the completion of these steps, the vision of your barndominium has taken a significant leap towards reality. Each beam, each wall, and each piece of tile placed, is a testament to your journey from a simple dream to the verge of a tangible, touchable, livable home. As you survey the progress, let the anticipation of the next phase fill you with excitement. The journey continues, and the best is yet to come.

Chapter 8

Estimating Your Costs: The Backbone of Your Barndominium Budget

In the world of barndominium building, budgeting is akin to the steady rhythm of a metronome, providing a constant beat guiding the symphony of construction. It's an instrumental part of the process that ensures your dream home doesn't turn into a financial nightmare. In this chapter, we'll take a deep dive into the process of creating a comprehensive budget for your barndominium building project.

Creating a Comprehensive Budget

Land Acquisition Costs

Let's start with the first and most fundamental cost - land acquisition. Purchasing the perfect plot of land for your barndominium is a significant investment, and the cost can vary widely depending on various factors like location, size, accessibility, and the presence of utilities.

It's like shopping for a designer handbag - you can find options at all price points, but it's important to find one that fits your budget and meets your needs. Consider not just the upfront purchase price but also any additional costs like closing fees, land surveys, or soil testing.

Material Costs

Next up on our budgeting list is the cost of building materials. From the concrete for your foundation to the shingles for your roof, every nut, bolt, and nail comes with a price tag.

Think about it when you're planning a dinner party. You need to account for every ingredient for each dish on your menu. Similarly, you need to list down every material you'll need for your barndominium and research their costs.

Remember, prices can vary based on factors like quality, brand, and location. It's a good idea to get quotes from multiple suppliers to ensure you get the best deal.

Labor Costs

Unless you're planning to do all the work yourself, labor costs will be a significant part of your budget. This includes wages for contractors, plumbers, electricians, and any other professionals you'll need to hire.

Imagine you're producing a movie. You have to pay each actor, cameraman, and crew member involved in making your vision come to life. In the same way, each professional involved in building your barndominium needs to be compensated for their skills and expertise.

When estimating labor costs, consider factors like hourly rates, the complexity of the job, and the estimated duration of the work. Always ask for detailed quotes to avoid any surprises down the line.

Permit and Inspection Fees

Building a barndominium isn't just about the physical construction process. There's also a maze of permits and inspections to navigate, and these come with their own set of fees.

Just like applying for a passport or a driver's license, these permits and inspections are necessary to ensure your barndominium meets all legal and safety standards. Fees can vary based on your local regulations, so it's important to check with your local building department to get an accurate estimate.

Utility Connection Charges

Finally, don't forget to factor in the cost of connecting utilities. Whether it's water, electricity, or internet, setting up these services often involves connection fees.

Think of it as the delivery charge when you're ordering a pizza. It's an additional cost on top of the pizza itself, but it's necessary to get the pizza delivered to your doorstep. Similarly, utility connection charges are necessary to get essential services delivered to your barndominium.

In conclusion, creating a comprehensive budget for your barndominium is a detailed and meticulous process. It's about accounting for every cost, big or small, and planning your spending in a way that aligns with your financial capabilities. With a well-planned budget in place, you can confidently stride forward in your barndominium building journey, knowing your dream home will also be a financially sound investment.

Predicting Unexpected Expenses

Weather-Related Delays

Just as a sudden downpour can disrupt a perfectly planned picnic, unforeseen weather conditions can throw a wrench into your barndominium building schedule. Torrential rains can make the site inaccessible, high winds can make it unsafe to erect walls or install the roof, and freezing temperatures can cause delays in concrete work. While we can't control the weather, we can certainly factor in the potential for weather-related delays in our budget.

To do this, consider the typical weather patterns in your area during the construction period. Are heavy rains or snowstorms common? Is it a hurricane-prone zone? Incorporate a buffer in your budget for potential weather-related delays. This buffer could cover additional labor costs, equipment rental extensions, or even the costs of securing the site and materials during inclement weather.

Changes in Material Prices

Building a barndominium is not an overnight task. It's a process that spans several months or even years. During this time, the prices of building materials can fluctuate due to factors like changes in supply and demand, import tariffs, or even global events. For instance, the price of lumber can skyrocket during a construction boom, or the cost of steel might increase due to new trade tariffs.

To protect your budget from these fluctuations, it's wise to include a contingency for changes in material prices. One approach is to secure price locks with your suppliers. This means the supplier guarantees the price for a certain period, protecting you from any price increases during that time. If this isn't possible, consider adding a 10-20% contingency to your materials budget to cover potential price increases.

Unforeseen Site Issues

While thorough site surveys and soil tests can reveal a lot about your land, there's always a chance of encountering unforeseen site issues once construction begins. You might discover rocky soil that's difficult to excavate, or you might hit groundwater when digging your foundation. Perhaps, the soil isn't as stable as you thought, requiring additional work to strengthen the foundation.

Dealing with these unexpected issues can add to your construction costs. Therefore, it's prudent to allocate a portion of your budget for such contingencies. The exact amount will depend on the potential risk factors associated with your site, but a good rule of thumb is to set aside 5-10% of your construction costs for unforeseen site issues.

Additional Design Changes

Imagine you're halfway through baking a cake, and you decide to tweak the recipe. Perhaps you want to add some nuts for extra crunch, or maybe you decide to make it a two-layer cake instead of just one. These last-minute changes will require additional ingredients, more baking time, and possibly even new kitchen tools.

Similarly, making changes to your barndominium design during construction, whether it's adding another window, moving a wall, or upgrading the kitchen fixtures, will add to your costs. While it's natural to want to tweak the design as your barndominium takes shape, it's important to understand the financial implications of these changes.

To account for this, consider setting up a separate budget for design changes. This will give you the flexibility to make adjustments along the way, without blowing your overall budget. Just remember to consult with your builder about the cost implications before you decide to add that extra skylight or move that bathroom wall.

In essence, budgeting for a barndominium is not a one-time task, but an ongoing process of monitoring, adjusting, and planning. It's about being prepared for the unexpected, and having the flexibility to adapt to changes, all while keeping a close eye on the bottom line. By predicting potential unexpected expenses and planning for them in your budget, you'll be better equipped to handle whatever surprises may come your way, ensuring a smoother and more enjoyable building experience.

Controlling Costs During Construction

Regular Monitoring of Expenses

Picture yourself as a ship's captain, navigating the turbulent waters of the construction process. Your budget is the compass guiding your ship, helping you stay on course amidst the waves of expenses. Regularly monitoring your expenses is like frequently checking your compass, ensuring you're on the right path.

To do this, set up a simple tracking system to record each expense as it arises. This could be a spreadsheet, a financial software program, or even a good old-fashioned ledger book. The key is to keep it updated religiously, recording each expenditure, no matter how small.

Furthermore, regularly compare your actual expenses with your budgeted amounts. Are you on track, or are you veering off course? This practice will help you identify any cost overruns early, allowing you to adjust your course before it's too late.

Efficient Use of Materials

In your quest to control costs, consider the materials you're using. Are you making the most efficient use of them? Wasted materials are like leaks in a ship, causing your budget to slowly sink.

Plan your material use carefully to minimize waste. For instance, when ordering lumber, try to utilize standard lengths to reduce offcuts. When installing drywall or flooring, plan your layout to make the most efficient use of each sheet or tile.

Even small efficiencies can add up to significant savings. It's like patching up those leaks in your ship, helping you keep your budget afloat and steer your project toward financial success.

Hiring Reliable Contractors

The crew on your ship plays a critical role in your voyage. A skilled and reliable crew can navigate through storms, fix leaks, and keep your ship sailing smoothly. Similarly, reliable contractors can help you control costs during the construction of your barndominium.

A reliable contractor will not only deliver quality work but will also stick to the agreed timeline and budget. They can offer valuable advice on cost-saving strategies, help you avoid costly mistakes, and ensure the job is done right the first time, saving you the cost of repairs or rework.

When hiring contractors, don't automatically go for the cheapest quote. Consider their reputation, experience, and the quality of their past work. Remember, a slightly higher expense upfront can result in significant savings down the line.

Avoiding Unnecessary Changes

As you sail towards the horizon, it can be tempting to change course, drawn by the allure of a distant island or the promise of smoother seas. But remember, every change of course requires extra time and fuel. Similarly, making changes during the construction process can add to your costs.

While some changes might be necessary due to unforeseen circumstances, try to avoid unnecessary changes that are driven by impulse or indecision. Each change can result in additional labor costs, material waste, and potential delays.

Before making a change, consider its impact on your budget and timeline. Is it worth the extra cost and potential delay? If not, it might be best to stick to your original plan.

In conclusion, controlling costs during construction is like steering a ship on a turbulent sea. It requires constant attention, careful decision-making, and efficient use of resources. But with regular monitoring of expenses, efficient use of materials, reliable contractors, and a disciplined approach to changes, you can navigate these waters successfully, keeping your budget shipshape and your barndominium dream on course.

Adjusting Your Budget as Needed

Re-evaluating Design Choices

Building your barndominium is like composing a melody. As the notes begin to form a harmonious tune, there might be instances where a certain note doesn't strike the right chord. This is similar to the need for re-evaluating design choices during the construction process.

Perhaps the custom-made kitchen cabinets you initially opted for are stretching your budget too thin. Or the high-end light fixtures you picked for the living room are no longer viable. In such scenarios, consider alternatives that offer similar functionality and aesthetics at a more affordable price. After all, the beauty of a barndominium lies in its adaptability and the scope for personalization.

Prioritizing Essential Costs

Think of your barndominium budget as a pie. While it would be great to have equal slices of all your favorite flavors, sometimes you have to make tough choices. This is where prioritizing essential costs comes in.

Primarily, focus on the elements that contribute to the structural integrity and longevity of your barndominium. These include a solid foundation, quality roofing, and efficient insulation. Aesthetics, while important, should be secondary to these essential costs. Remember, you can always upgrade aesthetic elements in the future as your budget permits.

Allocating Contingency Funds

In the realm of barndominium construction, surprises are not always pleasant. Unforeseen site issues, price escalations, or design changes can lead to additional expenses. To cushion the impact of these unexpected costs, it's wise to set aside a certain percentage of your budget as contingency funds.

Imagine you're packing for a camping trip. You'd pack an extra set of clothes and some additional food, just in case. Similarly, a contingency fund is like a safety net, catching you when unexpected costs threaten to throw you off balance.

Re-negotiating Contractor Rates

Finally, don't shy away from re-negotiating contractor rates if the need arises. Think of it as haggling at a flea market. If you feel the quoted price is not justified or if your budget is strained, open a dialogue with your contractor.

Perhaps they can suggest cost-cutting strategies, like using different materials or modifying the design. Or they might be willing to offer a discount on their labor rates. Remember, communication is key. A candid discussion about your budget constraints can lead to mutually agreeable solutions.

As you navigate the financial aspects of building your barndominium, remember that every dollar spent is a step towards realizing your dream home. By adjusting your budget as needed, you're not compromising on your dream, but simply steering it in a direction that leads to financial comfort and peace of mind. As you close your ledger and look around at the progress you've made, take a moment to appreciate the value of careful financial planning. It's not just about counting pennies, but about making each penny count. As we move forward, we'll explore another crucial aspect of building your barndominium - financing your project.

Chapter 9

Financing Your Barndominium: A Guide to Making Your Dream Home a Reality

Imagine standing on the brink of a vast ocean, ready to set sail on an epic adventure. You've charted your course and prepared your ship, but there's one critical element missing - the wind in your sails. In the journey of building your barndominium, financing is that wind. It's the driving force that propels your dream from the realm of imagination into tangible reality.

However, navigating the world of financial options can be as complex as sailing through a storm. Fear not, for this chapter is designed to be your trusty compass, guiding you through the intricacies of financing your barndominium. We'll examine various options, weigh their pros and cons, and provide you with the knowledge you need to make informed decisions. So, let's hoist the sails and catch the wind of financing.

Exploring Financing Options

The world of finance is a vast ocean, teeming with potential sources of funds for your barndominium project. Let's dive in and explore some of these options.

Personal Savings

One of the most straightforward ways to finance your project is through personal savings. This is akin to using the wind in your sails to propel your ship forward. It's a journey that requires careful planning and discipline, but it gives you full control over your finances.

Using personal savings to fund your project means you don't have to worry about interest rates, loan approvals, or repayments. However, it also means you need to have a substantial amount of money set aside specifically for your barndominium build. If you choose this route, ensure you still have enough savings left for emergencies or unexpected costs.

Home Equity Loans

If you're a homeowner with a significant amount of equity in your home, a home equity loan could be a viable financing option. Think of it as harnessing the power of the currents beneath your ship to help propel you forward.

Home equity loans allow you to borrow against the value of your home. They usually offer lower interest rates than personal loans or credit cards, making them a cost-effective way to finance your barndominium. However, it's important to remember that your home is used as collateral, meaning it could be at risk if you fail to make your repayments.

Construction Loans

Navigating the high seas of construction can be a daunting task, but with a construction loan, you have a sturdy ship to sail you through. Construction loans are short-term loans specifically designed to cover the cost of building a home.

These loans work on a draw-down basis, meaning the funds are released in stages as the construction progresses. This can be particularly beneficial as you only pay interest on the funds you've drawn down, not the total loan amount.

One thing to bear in mind, however, is that construction loans usually have higher interest rates and stricter approval requirements compared to other types of loans. Once the construction is complete, the loan is typically converted into a regular mortgage.

Government Grants and Subsidies

Finally, don't overlook the possibility of government grants and subsidies. Consider these as favorable winds, helping to propel your ship toward its destination.

Many governments offer grants or subsidies for new home constructions, especially if they meet certain criteria such as energy efficiency or use of sustainable materials. These can significantly offset the cost of building your barndominium. Be sure to check with your local or federal housing agencies to see what assistance may be available.

In essence, financing your barndominium is akin to charting a course across the vast ocean. There are numerous routes you can take, each with its challenges and rewards. By carefully considering each option and aligning it with your financial situation, you can ensure a smooth and successful voyage to the land of your dream home.

Applying for a Construction Loan

Preparing a Detailed Project Plan

Crafting a detailed project plan for your barndominium build can be compared to sketching a detailed map of a treasure island. It's a key document that outlines all aspects of your construction project, providing lenders with a clear vision of your project.

Your project plan should include specifics about the property, including its location and size. It should detail the design of your barndominium, including floor plans and exterior renderings. It's also important to outline your construction timeline, including key milestones such as when the foundation will be laid, when framing will be completed, and when the final finishes will be installed.

In addition, provide a detailed budget, breaking down the costs for each phase of construction. This should include estimates for labor costs, material costs, permit fees, and any other expenses related to the project.

Gathering Required Documentation

Assembling the necessary documentation for a construction loan is akin to packing your suitcase for a long trip. It's a critical step in the loan application process, requiring you to gather a variety of documents that lenders will use to assess your financial situation and the viability of your project.

Typically, lenders will ask for your credit report, proof of income, and a record of your assets and debts. They will also require a copy of your detailed project plan, complete with budget, timeline, and designs.

If you've hired a builder or contractor, you'll need to provide their details and credentials. This might include their license number, proof of insurance, and references from previous clients. If you're acting as your own general contractor, be prepared to demonstrate your ability to manage a project of this scale.

Comparing Loan Terms

Once you've prepared your project plan and gathered your documentation, it's time to start shopping around for a loan. This process is like visiting different markets to find the best quality ingredients for a feast.

Not all construction loans are created equal, and it's important to compare the terms of different loans to find the best fit for your needs. Key terms to consider include the loan term (how long you have to repay the loan), the interest rate, and whether the rate is fixed or variable.

Also, consider the draw schedule, which is how the funds will be disbursed as your project progresses, and what milestones you need to reach to receive each draw.

Understanding Interest Rates and Fees

The final piece of the puzzle in applying for a construction loan is understanding the interest rates and fees associated with the loan. This is similar to understanding the rules of a game before you start playing.

Interest rates on construction loans are typically variable, meaning they can fluctuate over the term of the loan. They are often higher than the rates on traditional mortgages, due to the higher risk associated with construction projects. It's important to factor the interest payments into your project budget, as you'll need to start making payments once the first draw has been made.

In addition to interest rates, there are also various fees associated with construction loans. These may include processing fees, inspection fees, and closing costs. Be sure to ask potential lenders for a detailed list of fees, so you can factor these into your budget and avoid any unpleasant surprises.

In essence, applying for a construction loan is a meticulous process that requires careful planning, thorough documentation, and a deep understanding of loan terms, interest rates, and fees. However, with a well-prepared project plan, a complete set of documents, and a clear understanding of your financial commitments, you can secure the funding needed to transform your barndominium dream into a reality. It's not just about crunching numbers, but about strategizing, planning, and making informed decisions that will set the foundation for your dream home.

Understanding Mortgage Options

Fixed-Rate Mortgages

Imagine buying a ticket for a long, exciting roller coaster ride, only to discover that the ride is smooth, with no unexpected dips or turns. That's the essence of a fixed-rate mortgage. It's a home loan with an interest rate that remains unchanged for the life of the loan.

Whether it's a 15-year or a 30-year term, the interest rate you lock in at the beginning is the rate you'll pay until your final payment. This unchanging nature makes budgeting easier, as your repayment amount is the same each month. However, the tradeoff is that fixed-rate mortgages often start with higher interest rates compared to other types of loans.

Adjustable-Rate Mortgages

Now, consider a different kind of roller coaster ride, one filled with exhilarating twists and turns. That's an adjustable-rate mortgage (ARM) for you. Unlike their fixed-rate counterparts, ARMs have interest rates that adjust over time.

Typically, ARMs start with a lower interest rate than fixed-rate mortgages, making them appealing for short-term homeowners or those expecting a future income hike. However, the fluctuating nature of the rates means your monthly payment can change, sometimes significantly. If you opt for an ARM, be prepared for potential increases in your monthly payment.

Government-Insured Loans

Picture a safety net below your tightrope walk, ready to catch you if you lose your balance. That's what government-insured loans are like. These are mortgages backed by the government, designed to help homebuyers who may not qualify for conventional loans.

The Federal Housing Administration (FHA), the Department of Veterans Affairs (VA), and the United States Department of Agriculture (USDA) are the main agencies offering these types of loans. FHA loans are popular among first-time home buyers due to their lower down payment requirements. VA loans are available for veterans and offer benefits like no down payment and no mortgage insurance. USDA loans aim to boost homeownership in rural and suburban areas and also offer zero down payment options.

Jumbo Mortgages

Finally, let's talk about jumbo mortgages, the giants of the mortgage world. These are loans that exceed the conforming loan limits set by Fannie Mae and Freddie Mac, the two government-sponsored entities that buy mortgages from lenders.

Jumbo mortgages are used to finance luxury properties and homes in highly competitive local real estate markets. Because of the larger amount, they come with stricter requirements in terms of credit score and debt-to-income ratio. They also often require larger down payments compared to conventional loans.

In essence, selecting a mortgage is like choosing a path to reach your destination. Each path has its unique terrain and scenery. Understanding the various mortgage options enables you to choose a path that complements your financial landscape and aligns with your homeownership goals. So, as you weigh your options and prepare to make your choice, remember that the right mortgage is the one that brings you comfortably and confidently closer to the doorstep of your dream barndominium.

Managing Loan Repayments

Setting Up Automatic Payments

With the excitement of your new barndominium project, it's easy to overlook the mundane task of managing your loan repayments. However, it's as crucial as keeping a keen eye on your compass while navigating unfamiliar waters. One approach to streamline this process is by setting up automatic payments.

Imagine a well-tuned clock, its gears moving in perfect harmony, keeping time without fail. Automatic payments work similarly. Once set up through your bank, they ensure your monthly loan payments are made on time, every time, without you needing to lift a finger. This not only saves you the hassle of remembering due dates but also helps maintain a good credit score by preventing missed payments.

Making Extra Payments

As your barndominium starts taking shape, you might find yourself with some extra funds - perhaps due to cost savings from efficient material use or lower-than-expected labor costs. A smart way to use these funds is by making extra payments towards your loan.

Think of these extra payments as tailwinds, helping you sail faster toward your financial goal. Making extra payments reduces the principal amount of your loan, which can significantly reduce the total interest payable and shorten your loan tenure. Check with your lender about the possibility and implications of making extra payments to ensure you're making an informed decision.

Refinancing Options

As you navigate the financial seas of your barndominium project, you may encounter shifting currents. Changes in personal finances, market conditions, or interest rates may prompt you to consider refinancing your loan.

Refinancing is akin to adjusting your sails to catch a more favorable wind. It involves replacing your existing loan with a new one, ideally with better terms such as a lower interest rate or a more manageable repayment schedule. However, refinancing comes with its costs, such as application fees, so it's important to weigh these against the potential benefits before deciding.

Dealing with Financial Hardships

Despite your best planning and budgeting efforts, you may encounter stormy weather in the form of financial hardships. These could be due to personal emergencies, unexpected expenses, or economic downturns. It's important to remember that such financial storms, while challenging, are not insurmountable.

If faced with financial hardships, the first step is to reach out to your lender. Most lenders have hardship programs in place and can offer solutions such as temporary payment reductions, loan modifications, or forbearance. While these measures can provide short-term relief, it's crucial to understand the long-term implications and make a plan to get back on track as soon as possible.

In essence, managing loan repayments is an integral part of your barndominium financing strategy. It requires diligence, adaptability, and proactive communication with your lender. By setting up automatic payments, making extra payments when possible, considering refinancing options, and having a plan for potential financial hardships, you can navigate the financial seas with confidence, steering your dream barndominium project towards a successful and financially sound completion.

As we wrap up this chapter, let's take a moment to appreciate the progress we've made. From exploring financing options to understanding mortgage options, from applying for a construction loan to managing repayments, we've navigated some complex waters. Now that we've charted our financial course, it's time to set sail towards the final stages of our barndominium building adventure. In the next chapter, we will delve into the delightful details of making your barndominium a true home.

Chapter 10

Making Your Barndominium a Home: The Art of Personalizing Your Space

Picture this. You've just finished a challenging puzzle. The final piece clicks into place, and you lean back, taking in the completed image. There's a sense of satisfaction, of accomplishment. Yet, the image is not your own. It's a pre-designed pattern, a preset end goal. Now, imagine being given a blank canvas, an array of colors, and the freedom to paint whatever you wish. That's what furnishing your barndominium feels like. It's your chance to infuse your personality into the fabric of your home, creating an environment that is uniquely, authentically you.

Your barndominium, with its open layout and rustic charm, is a perfect canvas for your creativity. In this chapter, we'll explore how to select furniture, choose durable materials, and incorporate rustic elements, transforming your barndominium into a personalized reflection of your style and values.

Furnishing Your Space

Selecting Furniture That Complements the Open Layout

When you walk into a beautifully designed restaurant, what's the first thing you notice? It's not the individual tables or chairs, but the overall setup - the way the furniture is arranged to create a welcoming, cohesive environment. The same principle applies when furnishing your barndominium.

With its open layout, your barndominium offers unique opportunities for furniture placement. Instead of thinking in terms of separate rooms, consider your entire living area as a single, unified space.

For instance, you might position your sofa and coffee table to face a stunning countryside view, creating a visual anchor for your living area. Then, arrange your dining table and chairs nearby, but angled differently to delineate the dining space without breaking the visual flow.

In this scenario, your furniture serves a dual purpose. It not only provides comfort and functionality but also helps define different zones within your open layout. As you select and arrange your furniture, think of yourself as a sculptor, using each piece to shape and define your living space.

Choosing Durable Materials for High-Traffic Areas

Have you ever seen a well-trodden path in a park or a forest? It's usually packed hard, showing signs of constant use. Your barndominium will also have well-trodden paths - areas that see high traffic like the entrance, living room, and kitchen.

Just like the hard-packed soil on a forest path, the furniture in these areas needs to be durable to withstand constant use. When shopping for furniture, don't just consider the style and price. Pay attention to the materials and construction.

Hardwood furniture, for example, is known for its durability and can withstand heavy use. Fabrics like leather or synthetic microfiber are also excellent choices for upholstery, as they are resistant to wear and tear and easy to clean.

Remember, sometimes investing in high-quality, durable furniture can save you money in the long run by reducing the need for frequent replacements.

Incorporating Rustic Elements to Maintain the Barn Aesthetic

One of the unique aspects of a barndominium is its rustic charm. The blend of simple, rugged construction with modern comforts creates a distinct aesthetic that sets barndominiums apart from traditional homes.

When furnishing your barndominium, consider incorporating rustic elements to maintain this aesthetic. This could mean choosing furniture made of reclaimed wood, adding a vintage leather armchair, or even using an old wooden crate as a side table.

Additionally, consider the color palette. Earthy tones like browns, greens, and grays can help enhance the rustic feel, while a pop of color can add a modern twist.

Remember, it's not about replicating a barn, but about creating a balance between the rustic and the refined. It's about embracing the barndominium aesthetic and adding your unique touch to it.

Furnishing your barndominium is more than just filling it with furniture. It's about creating a space that reflects your style, caters to your needs, and feels like home. It's about seeing your barndominium not just as a building, but as a canvas for your creativity, a testament to your taste. So, as you select your furniture, choose your materials, and incorporate rustic elements, remember that you're not just decorating a house. You're creating a home that is uniquely, authentically you.

Landscaping and Outdoor Living

Designing a Garden That Blends With the Natural Surroundings

Imagine a beautiful painting; its colors and textures flow seamlessly, creating a harmonious masterpiece. Similarly, when designing a garden for your barndominium, the goal is to create a natural flow between your home and the surrounding landscape.

Start by observing your property through different seasons. Notice how the light changes, which areas get the most sun or shade, and how the wind moves through your property. These observations will guide your decisions about where to plant trees, create sitting areas, or install water features.

Choosing native plants can enhance the natural feel of your garden. They typically require less maintenance and are more resistant to pests and diseases. Plus, they contribute to the local ecosystem, providing food and habitat for native birds and insects.

Building an Outdoor Entertainment Area

The beauty of living in a barndominium is the opportunity to blur the line between indoor and outdoor living. An outdoor entertainment area can serve as an extension of your living space, perfect for BBQs, family gatherings, or lazy Sunday afternoons.

Consider adding a patio or deck just outside your main living area. Furnish it with comfortable seating and a dining table for alfresco meals. If space allows, you might even install an outdoor kitchen, complete with a grill and a sink for easy meal preparation and cleanup.

Remember to consider privacy and shelter in your design. A pergola or a shade sail can provide protection from the sun, while fences or tall plants can create a sense of seclusion even if your neighbors are nearby.

Installing a Fire Pit for Cozy Gatherings

There's something magical about gathering around a fire. The warmth, the flickering light, and the crackling sounds create a cozy and inviting atmosphere. Installing a fire pit in your garden can provide a focal point for outdoor gatherings.

Fire pits come in many designs to suit your style and budget. A simple, portable fire pit can provide flexibility, or you might choose to build a permanent one using bricks or stones. Consider adding a seating area around the fire pit, with comfortable chairs or even built-in benches.

Remember to install your fire pit safely away from trees, fences, and other structures to prevent fire hazards. Also, check your local regulations concerning open fires to ensure you comply.

Creating a Vegetable Garden for Sustainable Living

Growing your own vegetables can be a rewarding and practical way to utilize your outdoor space. Not only does it provide fresh, organic produce for your meals, but it also offers a therapeutic hobby.

Start small, with easy-to-grow vegetables like lettuce, tomatoes, or herbs. As you gain confidence, you can expand your garden to include a wider variety of produce. Consider using raised beds to make planting, weeding, and harvesting easier on your back.

A well-planned vegetable garden can be aesthetically pleasing as well as productive. With some creativity, you can design beautiful patterns with different plants, turning your vegetable garden into a living work of art.

In essence, landscaping and outdoor living are about creating connections - with nature, with your family and friends, and with the food you eat. It's about making the most of your outdoor space and enhancing the quality of life in your barndominium. So, as you design your garden, build your entertainment area, install your fire pit, and plant your vegetable garden, remember that each element is a thread that weaves the tapestry of your barndominium lifestyle. It's not just about building a home, but about creating a lifestyle that reflects your values and enriches your life.

Maintaining Your Property

Regular Inspection of the Metal Structure for Rust or Damage

Think of your barndominium as a trusted vehicle, one that calls for regular check-ups to ensure all parts are functioning optimally and to detect any potential issues early. The metal structure of your barndominium, being a significant component of your home, requires routine inspections.

Just as you'd inspect a car for rust or damage to maintain its performance and appearance, you should routinely examine the metal elements of your barndominium. Look for any signs of rust or corrosion, particularly in areas exposed to moisture. Also, check for any signs of physical damage like dents or scratches, which could potentially lead to rust if left untreated.

Use a rust-preventive primer and rust-resistant paint to treat any areas showing signs of rust. If you notice significant damage or rusting, consider consulting a professional for advice on repair or replacement.

Cleaning and Sealing Wooden Elements

Wooden elements in your barndominium add a touch of rustic charm and warmth to your home. However, like any natural material, wood requires care to maintain its beauty and longevity.

Think of it as tending to a beautiful wooden antique. It needs regular cleaning to remove dust and grime, and periodic sealing to protect it from moisture and pests.

For cleaning, use a soft cloth and a gentle, wood-safe cleaner. Avoid using excessive water, as it can seep into the wood and cause damage. For sealing, use a high-quality wood sealer and apply it according to the manufacturer's instructions, paying special attention to outdoor wooden elements that are exposed to the elements.

Scheduling Professional Inspections for Plumbing and Electrical Systems

Your barndominium's plumbing and electrical systems are like the veins and nerves of your home, quietly working behind the scenes to provide water, warmth, and light. To ensure they continue to function efficiently and safely, regular professional inspections are crucial.

Consider it as a wellness check-up from a specialist doctor. An experienced plumber can check your pipes for leaks, examine the water pressure, and ensure all fixtures are functioning correctly. Similarly, a licensed electrician can inspect your wiring, test your outlets, and ensure your home is safe from electrical hazards.

It's advisable to schedule these inspections annually, even if everything seems to be working fine. It's a proactive measure that can help detect and address potential issues before they become costly problems.

Implementing a Pest Control Plan

Living in a rural setting can sometimes mean sharing your surroundings with various critters and insects. While some are harmless, others can pose a threat to your barndominium. Implementing a pest control plan is like setting up a security system for your home, keeping unwelcome intruders at bay.

Start by sealing any gaps or cracks in your walls, doors, and windows where pests could enter. Keep your kitchen and dining area clean and store food in sealed containers to avoid attracting pests. Regularly empty your garbage and ensure your compost heap, if you have one, is well away from your home.

For larger pests like rodents or raccoons, consider installing a fence around your property. Remember, dealing with a minor pest problem promptly can prevent it from becoming a major infestation. If you notice signs of pests, such as droppings, nests, or damage to your property, consider hiring a professional pest control service.

The process of maintaining your property is akin to taking care of a cherished garden. It requires regular attention, occasional intervention, and a commitment to preserving its beauty and function. With each inspection, each cleaning session, and each preventive measure, you're not just maintaining a structure; you're nurturing a home that shelters, comforts, and reflects your lifestyle.

Embracing the Barndominium Lifestyle

Hosting Barn Parties and Community Events

There's a unique charm to barn parties that is hard to replicate in a traditional setting. The open layout of your barndominium, coupled with the rustic aesthetics, creates the perfect backdrop for a variety of events. From casual BBQs and birthday parties to more formal gatherings like weddings or community events, your barndominium can play host to memorable occasions.

Imagine a summer evening, the aroma of grilling food wafting through the air as friends and family gather around, their laughter echoing through the open space of your barndominium. Or picture a winter night, the warmth of a roaring fire drawing people closer, as the walls of your barndominium reverberate with the melodies of holiday songs. By hosting events, you're not just making use of your space; you're creating memories that will be cherished by your loved ones.

Embracing the DIY Spirit for Home Improvements

One of the joys of owning a barndominium is the opportunity to put your stamp on it. The straightforward construction and open layout make it an ideal canvas for DIY enthusiasts. From building custom furniture and installing light fixtures to painting walls and laying tiles, there are ample opportunities to roll up your sleeves and get involved.

Imagine the satisfaction of sitting on a deck that you built with your own hands or the pride you feel when guests compliment the kitchen backsplash you installed. Embracing the DIY spirit not only allows for personalized home improvements but also provides a sense of accomplishment and ownership. Plus, it's a great way to save money and learn new skills.

Enjoying the Tranquility and Privacy of Rural Living

Perhaps one of the most significant benefits of the barndominium lifestyle is the ability to enjoy the tranquility and privacy that rural living offers. Away from the hustle and bustle of city life, you can enjoy the peace of quiet mornings, the spectacle of starlit nights, and the freedom of wide-open spaces.

Picture yourself sipping coffee on your porch, watching the sun rise over a panorama of fields and forests. Or consider the serenity of a quiet evening, the only sounds being the soft whisper of the wind and the distant hoot of an owl. This tranquility can offer a respite from the hectic pace of modern life, providing a space for relaxation and rejuvenation.

Living in a barndominium offers an experience that is as unique as it is enchanting. It's more than just a type of dwelling; it's a lifestyle, a statement of personal values and preferences. As you host parties, embark on DIY projects, and bask in the tranquility of rural living, you're not simply occupying a space; you're living your dream, one beautiful day at a time.

As we close this chapter, we open the door to the final stages of your barndominium journey. We've covered a lot of ground, and the finish line is in sight. But before we cross it, there's one more important topic to discuss - understanding the potential resale value of your barndominium. So, let's turn the page and delve into the final chapter of this exciting adventure.

Chapter 11

The Potential Resale Value: Unfolding the Future of Your Barndominium Investment

Exhale. Close your eyes, and picture a bustling marketplace. The chatter of eager buyers, the rustle of crisp bills changing hands, the satisfying thump of a gavel confirming a successful sale. This scene is not far removed from the real estate market, a dynamic arena where properties change hands, dreams are bought and sold. As a future barndominium owner, this market holds significant relevance for you. It's the sphere where your property might one day find a new owner, where the financial returns on your investment will be realized.

In this chapter, we'll step into this marketplace, not as buyers, but as potential sellers. We'll learn how to understand the market for barndominiums, analyze recent sales, identify the unique selling points of our property, and keep up with the latest trends. It's about peering into the future of your investment, preparing for the day when you might decide to pass on the keys of your beloved barndominium to a new owner.

Understanding the Market for Barndominiums

Analyzing Recent Sales of Barndominiums in Your Area

Think of this as a detective mission, where your clues are found in the pages of property listings and sale records.

Start your investigation by researching recent barndominium sales in your area. Local real estate websites, county property records, and even local newspapers can be valuable sources of information. Look out for details such as the selling price, the time on the market, and the features of the sold properties.

Identifying Unique Selling Points of Your Property

Now, turn your detective lens towards your property. What sets your barndominium apart? Perhaps it's the breathtaking view from the deck, the energy-efficient design, or the custom-built barn doors.

These unique features are your property's selling points. They are the elements that will capture potential buyers' interest and set your barndominium apart from others. Make a list of these features and consider how they might appeal to a potential buyer.

Keeping Up with Trends in the Barndominium Market

Finally, keep an eye on the evolving trends in the barndominium market. Just as fashion trends influence what we wear, real estate trends can affect property values and buyer preferences.

Stay informed about the latest designs, technologies, and amenities popular among barndominium owners. Regularly check home design websites, follow real estate blogs, and join barndominium communities on social media.

Keeping up with these trends will not only help you maintain and upgrade your property to stay competitive in the market, but it can also provide insights into what future buyers might be looking for in a barndominium.

Understanding the market for barndominiums is like learning the rules of a game. It equips you with the knowledge and insights needed to play the game effectively when the time comes. By analyzing recent sales, identifying your property's unique selling points, and keeping up with market trends, you're not just preparing for a potential future sale. You're ensuring that when the time comes to sell, your barndominium will stand out in the marketplace, its value recognized and appreciated by discerning buyers.

Enhancing the Resale Value

Investing in Energy-Efficient Upgrades

As we move towards a more sustainable future, energy efficiency has become a buzzword in real estate. Energy-efficient homes are not just good for the environment, they are also good for your wallet, offering significant savings on utility bills.

Imagine a well-insulated barndominium, its cozy interior holding the warmth during a chilly winter night, or keeping the cool air in during a sweltering summer day. This is the magic of energy-efficient upgrades, and it's a magic prospective buyers will be willing to pay for.

Consider installing energy-efficient appliances, like an ENERGY STAR refrigerator or dishwasher, that use less electricity without compromising on performance. Opt for LED light fixtures that consume less energy and last longer than traditional bulbs. Also, consider upgrading your insulation, windows, and doors to minimize heat loss and keep your home comfortable, regardless of the weather outside.

Adding Extra Amenities Like a Hot Tub or Outdoor Kitchen

The allure of a barndominium lies not just in its rustic charm and open layout, but also in the potential to create unique living spaces. Extra amenities like a hot tub or an outdoor kitchen can significantly enhance the appeal of your barndominium, making it stand out in the resale market.

Visualize a luxurious hot tub on your deck, the warm water inviting relaxation after a long day. Or imagine an outdoor kitchen, complete with a grill and a pizza oven, ready to host memorable summer BBQs. These amenities add to the lifestyle experience of living in a barndominium, making it more than just a home, but a personal retreat.

When adding these amenities, focus on quality and professional installation. Also, consider their placement for optimal functionality and aesthetics. For instance, a hot tub could be placed to offer a panoramic view of your property, while an outdoor kitchen should be conveniently located near the indoor kitchen for easy access to ingredients and utensils.

Maintaining the Property in Excellent Condition

Like a well-oiled machine, a well-maintained barndominium operates smoothly and efficiently. Regular maintenance not only ensures your home stays in top condition but also contributes to its resale value. After all, a well-cared-for home is more attractive to potential buyers, reducing the need for them to invest in repairs or upgrades.

Create a regular maintenance schedule that covers all aspects of your barndominium. This includes routine tasks like cleaning the gutters, inspecting the roof, testing the smoke detectors, and servicing the HVAC system. It also includes seasonal tasks like preparing your home for winter or doing a spring clean-up of your property.

Remember, maintenance is an ongoing task, but it pays off in the long run. A well-maintained barndominium not only provides a safe and comfortable living environment, but it also holds its value better in the resale market. As you sweep the floors, prune the trees, and repaint the walls, know that each task contributes to preserving the value of your investment, ensuring your barndominium continues to be a source of pride and joy.

Preparing for a Future Sale

Documenting All Improvements and Maintenance Work

Imagine your barndominium as a beautiful masterpiece that keeps evolving over time. Each stroke of the paintbrush, each tiny adjustment, adds to its overall charm and value. Similarly, every improvement or maintenance task you undertake in your home adds to its worth. However, unlike a painting, these enhancements aren't always visible to the naked eye. Hence, it's important to maintain a thorough record of all such activities.

Think of this as a detailed diary of your home. From minor tasks like annual HVAC servicing to major ones like installing a new roof, everything should be documented. Include the date of the work, the contractor details, and the cost incurred. Also, keep a copy of all relevant permits and inspection reports. This record not only serves as a testament to your diligent upkeep of the property but can also be a powerful tool during price negotiations with potential buyers.

Staging the Property to Highlight Its Best Features

Staging your property is akin to setting the stage for a grand performance. It's about showcasing your barndominium in the best light, highlighting its unique features, and creating a warm, inviting atmosphere that potential buyers can envision themselves in.

Start by decluttering your home. A clean, uncluttered space not only looks more appealing but also allows buyers to focus on the property itself rather than your personal belongings. Next, arrange the furniture to showcase the functionality and spaciousness of your open layout. Use rugs, lighting, and accessories to define different zones within your space.

Don't forget the outdoor areas. Mow the lawn, prune the trees, and clean the patio. If you have an outdoor kitchen or a fire pit, consider setting it up as if you were hosting a party. This not only highlights these amenities but also helps buyers visualize the lifestyle your barndominium offers.

Hiring a Professional Photographer for Listing Photos

In today's digital age, the first impression of your barndominium for most buyers will be through the listing photos. Hence, investing in professional photography can make a significant difference in attracting potential buyers.

A professional photographer understands how to use lighting, angles, and composition to capture the best features of your property. They can make your rooms look spacious and bright, highlight unique architectural details, and portray the beauty of your outdoor spaces.

Moreover, a good photographer can create a visual narrative that draws buyers in. They can capture the essence of the barndominium lifestyle, from the tranquility of rural living to the charm of the barn-inspired interiors, making your listing stand out in the bustling real estate market.

In essence, preparing your barndominium for a future sale is about attention to detail and thoughtful presentation. It's about keeping a meticulous record of your home's evolution, setting the stage for buyers to appreciate its unique features, and capturing its essence through professional photography. These preparations not only help attract potential buyers but also ensure you receive the best possible price for your cherished property. And while the thought of parting with your beloved barndominium might be bittersweet, knowing that you've prepared it well for its next owners can offer a sense of satisfaction and closure.

Selling Your Barndominium

Collaborating with a Realtor Experienced in Unique Properties

Selling a barndominium is not quite the same as selling a conventional home. The unique blend of rustic charm and modern amenities that define barndominiums requires a different selling approach. It's like selling a rare vintage car - you need someone who understands its unique appeal and can communicate that to potential buyers.

Working with a realtor who has experience selling unique properties can make a significant difference. They understand the niche market for barndominiums, know how to highlight its unique features and have the right contacts to reach potential buyers.

So, take your time in selecting a realtor. Look for someone who not only has a proven track record in selling unique properties but also shares your appreciation for the barndominium lifestyle. This synergy between you and your realtor can make the selling process smoother and more successful.

Attracting the Right Audience

The next step in selling your barndominium is to attract the right audience. Just as a talented musician performs best in front of an appreciative audience, your barndominium will sell better when presented to the right buyers.

Your target audience is likely to be people who appreciate the rural lifestyle, value unique home designs, and are drawn to the warmth and charm of a barndominium. They might be looking for a primary residence that offers peace and tranquility, a vacation home where they can escape from city life, or even a unique venue for their business.

To reach this audience, consider where they might be looking for properties. This could be on real estate websites that specialize in unique homes, at-home and garden shows, or in lifestyle magazines. Your realtor can also use their network to reach potential buyers.

Negotiating the Sale Price and Terms

Finally, once you've attracted a potential buyer, the last hurdle is to negotiate the sale price and terms. This is a delicate dance, a balancing act of securing the best price for your property while ensuring the buyer feels they're getting their money's worth.

Before entering negotiations, have a clear idea of your minimum acceptable price. Also, be prepared for potential buyers to negotiate not just the price, but also the terms of the sale. This could include things like the closing date, contingencies on selling their current home, or requests for certain repairs or improvements to be made before the sale.

Remember, negotiation is a two-way street. While you want to get the best price for your barndominium, the buyer also wants to feel they're making a good investment. With open communication, flexibility, and a bit of give-and-take, you can arrive at a sale price and terms that both you and the buyer are happy with.

Selling your barndominium is the final step in your journey as a barndominium owner. It's the point where you pass the baton to a new owner, someone who will hopefully cherish and enjoy the property as much as you have. And while it may be the end of your barndominium story, it's the start of a new chapter for the new owners. As for you, the knowledge, experiences, and memories you've gained from building and living in your barndominium are yours to keep, treasures that will last a lifetime.

And while the curtains close on this chapter, your journey continues. Life is an ongoing adventure, a series of chapters that make up the book of our lives. And who knows, perhaps in the next chapter, you'll find yourself embarking on another exciting adventure, building another dream, creating another masterpiece. After all, life is a canvas, and you hold the brush.

Conclusion

Reflect, Rejoice, and Inspire

As we close the final chapter of this comprehensive guide, let's pause for a moment and reflect on the incredible journey we've embarked on together. From envisioning your dream barndominium to witnessing it rise from the ground, every step, every decision, and every challenge faced has led you to a unique testament of your vision and tenacity - Your Barndominium.

Your barndominium is not just a structure made of concrete, metal, and wood. It's the embodiment of your dreams and desires. Each room, each feature, bears the imprint of your personality and style. As you walk through the spaces you've designed, pause and appreciate the journey that brought you here. The late-night planning, the meticulous budgeting, the thrill of seeing your design come to life — all these experiences are woven into the very fabric of your home.

Yet, the journey doesn't end here. As your life evolves, your barndominium will continue to grow and adapt with you. Perhaps a future improvement will be a new patio for summer BBQs or a cozy reading nook by the window. Maybe you'll add solar panels to harness the power of the sun or plant a beautiful rose garden. Your barndominium, with its flexible design and robust structure, is ready to accommodate these changes, mirroring the evolution of your lifestyle and aspirations.

Embrace the lifestyle your barndominium offers. Enjoy the tranquility of rural living, the comfort of modern amenities, and the satisfaction of dwelling in a home that truly reflects your personality. From hosting barn parties to enjoying quiet evenings by the fireplace, every moment spent in your barndominium is a celebration of your vision and hard work.

As you settle into this rewarding lifestyle, consider sharing your journey with others. Inspire those around you who dream of a unique, personalized dwelling. Share your experiences, the lessons learned, and the joy of seeing your dream materialize. Encourage them to take that first step towards their barndominium dream.

Building your barndominium has been an incredible adventure, one filled with challenges and victories, planning, and spontaneity. It's been a privilege to guide you through this journey, sharing insights, strategies, and the sheer excitement of creating something uniquely yours. As you step into this new chapter of your life, remember that your barndominium is more than a home. It's a testament to your vision, a symbol of your perseverance, and a canvas for your ongoing creativity.

Here's to the journeys yet to come, the dreams yet to be realized, and the endless possibilities that your barndominium offers. Enjoy every moment in the home you've crafted with love, patience, and an indomitable spirit. After all, the true joy of a barndominium is not just in building it, but in living it.

Testimonial from the Author:

My Barndominium journey began when my family bought a piece of land up north from our year-round home. It was a lot with septic and an electric hookup for an RV. We spent many weekends there, using our RV and enjoying the surrounding areas and lakes. It didn’t take long for me to start wondering and researching an alternative to RV vacationing. We intended to leave the RV on this lot, so, what if we had a cabin instead?

I have always been handy, making extra income on the side finishing basements into living spaces, remodeling bathrooms, and kitchens. Landscaping and other outdoor building jobs. I decided that I could take on a project such as building a cabin for our family vacation spot. I spent a spring brainstorming how I might want this cabin to take shape. How big would I want it to be? Do I want one floor or two? At every turn of this brainstorming, I was always landing on money as being a huge sticking point. In a traditional house or cabin, once you break ground, you are really under the gun to get things moving until the project is at least weathered in. In every scenario I could think of, foundation work was paramount to getting started. Being a family of 5, and living a typical middle-class lifestyle, I didn't have the money to simply contract out foundation work, or to hire a company to build a cabin. I knew though, that if I could get the shell of a building constructed, and within my budget, I could take on the rest. This is when the Barndominium idea was introduced to me. A co-worker mentioned this while he and I were talking during lunch. I hadn't heard of a Barndominium before he suggested it. When I started to research it, the idea grew on me. After countless bar napkins, loose-leaf sheets of paper, and scribbled-out graph paper designs, I was able to find a local contractor to build me a post-frame pole barn shell for my Barndo. Winter season came and went, and I was able to get a cement floor poured inside of my building. After I had walls and cement, I began framing out my design. Building the walls, installing insulation, plumbing, and electrical fixtures. The more we went up there to vacation, the more I was able to get finished. This process would have certainly been much faster had I the budget to contract out more of the work, but that just wasn't in the cards. Slowly coming together, with mistakes along the way, our Barndo took shape.

I enjoyed every aspect of my journey to realizing our family's Barndo. Once finished, I immediately began to scheme up building a larger version for us to move to permanently. This book, and the idea for the company, came together due to the lack of resources available for folks like me, who wanted to take on this journey. There are some books, videos, and resources out there. But, certainly plenty of room for much much more information. Everyone who sees what we have built draws interest, I answer questions daily on how we did this, or who thought of doing that. The local company that built the shell of my pole barn has been busy all around with families looking to do something quite similar.

I hope that you realize your dream home, whenever or however it might take shape. And I truly hope that BarnDream Builders can help inform you, guide you, and inspire you.

Thank you so much for taking the time to have a look.

BarnDream Builders

Congratulations!!!

You've reached the end of our book, "Building Your Barndominium: A Comprehensive Guide to DIY Barn Home Building, Creating Rural Living Home Designs, and Unique Barndo Spaces on Any Budget" by BarnDream Builders. We hope you're now armed with the knowledge, inspiration, and excitement to embark on your barndominium adventure.

But wait, there's one more important step on this journey. It's time to pay it forward and help others who are just starting or searching for guidance on their own barndominium projects.

Why Your Review Matters:

Your honest opinion can light the way for other potential barn home builders, showing them where they can find the valuable information and inspiration they need. By leaving a review on Amazon, you become a beacon of hope and knowledge for those following in your footsteps.

Passion is Contagious:

Building barndominiums is not just about construction; it's about building dreams, and your review can ignite the passion in others. When you share your experience and insights, you contribute to keeping the barn home dream alive.

Join the Movement:

By leaving a review, you become part of our community of builders who believe in sharing knowledge and supporting one another. Together, we can make a real difference in the lives of those aspiring to create their dream homes.

Thank You for Your Help:

We want to express our heartfelt gratitude for your willingness to help others on their barndominium journey. Your review is a gift that keeps on giving, and it's a powerful way to pay it forward.

Scan to Leave Your Review:

To leave your review on Amazon and become a guiding light for future barn home builders, simply click the link below:

Click here to leave your review on Amazon.

Thank you for being a part of this incredible journey. Together, we'll continue to build dreams and inspire others to do the same.

With immense gratitude,

BarnDream Builders

References

Barndominiums: Cost-Effective Building – BuildMax; *https://buildmax.com/cost-effective-building/#:~:text=Barndominiums%20often%20employ%20more%20cost,what%20you%20end%20up%20with*.

Barndominium Foundations: 5 Types To Know; *https://barndominiumzone.com/foundations/*

Barndominium Moisture Problems: A Comprehensive Guide; *https://barndominiumgold.com/barndominium-moisture-problems/*

Barndominium Resale Value Secrets: 6 Key Factors That ...; *https://barndominiums.co/barndominium-resale-value/*

Barndominium Summer Maintenance Checklist; *https://thebarndominiumcompany.com/barndominium-summer-maintenance-checklist/*

Best Flooring For Barndominiums; *https://www.barndominiumlife.com/best-flooring-for-barndominiums/*

Building Permit Process: A Complete Guide on All the Steps; *https://mykukun.com/blog/building-permit-process-guide/*

Building Your Dream Barndominium: A Step-by-Step Guide ...; *https://www.linkedin.com/pulse/building-your-dream-barndominium-step-by-step-guide-haven-mcclintic*

Construction Loans: What You Need To Know; *https://www.rocketmortgage.com/learn/construction-loans*

Construction Management: Understanding Cost Control; *https://www.fool.com/the-ascent/small-business/construction-management/articles/cost-control-in-construction/*

DIY Barndominiums: Tips for Choosing the Right Land to Build ...; *https://raisedinabarndominium.com/tips-for-choosing-the-right-land-to-build-on/*

Everything You Need to Know About Land Surveying; *https://mjslandsurvey.com/blog/guide-to-land-surveying/*

Financing a Barndominium - How to Find a Lender; *https://www.barndominiumlife.com/financing-a-barndominium-how-to-find-a-lender/*

Getting a Mortgage When Building Your Own Home; *https://www.investopedia.com/articles/personal-finance/032315/getting-mortgage-when-building-your-own-home.asp*

How Much Does a Barndominium Cost to Build? (2023); *https://homeguide.com/costs/barndominium-cost*

How Much Does It Cost to Build a House? *https://www.ramseysolutions.com/real-estate/how-much-does-it-cost-to-build-a-house*

How to Build Your Own House: A Step-by-Step Guide; *https://www.thespruce.com/building-your-own-house-1821301*

How to Frame a Barndominium: 8 Simple Steps; *https://www.barndominiumlife.com/how-to-frame-a-barndominium/*

Managing Unexpected Costs in Construction Projects; *https://linarc.com/buildspace/managing-unexpected-costs-in-construction-projects/*

Master the Art of Negotiating Land Sales with Confidence; *https://www.landbrokermls.com/blog/negotiating-a-land-sale/*

Metal Building Insulation: The 5 Best Options; *https://www.westernstatesmetalroofing.com/blog/metal-roofing-building-insulation*

Modern Barndominium: 7 Stunning Designs in 2022; *https://www.trulogsiding.com/barndominium-designs/*

Pole Barn Insulation: Your Options and Which is the Best ...; *https://sheafferconstruction.com/pole-barn-insulation/*

Popular Trends in Open Concept vs Traditional Layouts; *https://mybarndoplans.com/popular-trends-in-open-concept-vs-traditional-layouts/*

The Ins And Outs Of Barndominium Resale Value; *https://www.amandabrown.net/barndominium-resale-value/*

The Step by Step Process for Construction Site Preparation; *https://procrewschedule.com/the-step-by-step-process-for-construction-site-preparation/*

The Ultimate Guide to Barndominium Roof Styles; *https://www.barndominiumlife.com/guide-to-barndominium-roof-styles/*

Top 7 Benefits of Living in a Barndominium; *https://lynnbaber.com/top-7-benefits-of-living-in-a-barndominium/*

Top 10 Tips for Planning a Home Addition; *https://www.higgasonhomes.com/top-10-tips-planning-home-addition/*

What Upgrades Increase Home Value? 25 High-ROI ...; *https://www.homelight.com/blog/what-upgrades-increase-home-value/*

Your Complete Guide to Barndominium Requirements; *https://barndominiums.co/barndominium-requirements/*

Your Guide to Rustic Style Furniture; *https://www.ezmountain.com/blog/your-rustic-furniture-guide*

5 Construction Laws to Know Before You Build a House; *https://www.thisoldhouse.com/home-finances/21097121/5-common-construction-laws-you-should-know-before-you-build*

8 Amazing Barndominium Landscaping Ideas; *https://www.barndominiumlife.com/barndominium-landscaping-ideas/*

8 Creative Real Estate Marketing Ideas And Examples; *https://www.searchenginejournal.com/real-estate-marketing-creative-ideas-examples/470360/*

20 Tips for Repaying Your Home Loan Faster; *https://uno-webflow.webflow.io/articles/20-tips-repaying-home-loan-faster*